THE
# VORTEX
# ATOM
A New Paradigm

**Other Related Titles from World Scientific**

*Lectures on the Non-Equilibrium Theory of Condensed Matter*
*Second Edition*
by Ladislaus Alexander Bányai
ISBN: 978-981-122-380-8

*Lectures on Quantum Field Theory*
*Second Edition*
by Ashok Das
ISBN: 978-981-122-086-9
ISBN: 978-981-122-216-0 (pbk)

*Theory of Groups and Symmetries: Representations of Groups and*
*Lie Algebras, Applications*
by Alexey P Isaev and Valery A Rubakov
ISBN: 978-981-121-740-1

*The Quantum Puzzle: Critique of Quantum Theory and Electrodynamics*
by Barry R Clarke
ISBN: 978-981-4696-96-8
ISBN: 978-981-3226-01-2 (pbk)

# THE
# VORTEX
# ATOM
## A New Paradigm

## Barry R Clarke

Brunel University, UK

**World Scientific**

NEW JERSEY · LONDON · SINGAPORE · BEIJING · SHANGHAI · HONG KONG · TAIPEI · CHENNAI · TOKYO

*Published by*

World Scientific Publishing Co. Pte. Ltd.

5 Toh Tuck Link, Singapore 596224

*USA office:* 27 Warren Street, Suite 401-402, Hackensack, NJ 07601

*UK office:* 57 Shelton Street, Covent Garden, London WC2H 9HE


**Library of Congress Cataloging-in-Publication Data**
Names: Clarke, Barry R., author.
Title: The vortex atom : a new paradigm / Barry R Clarke.
Description: New Jersey : World Scientific, [2021] | Includes bibliographical references and index.
Identifiers: LCCN 2020053103 | ISBN 9789811227585 (hardcover)
Subjects: LCSH: Angular momentum (Nuclear physics) | Atomic hydrogen. | Atoms--Models. |
     Vortex theory (Astrophysics) | Hyperfine structure. | Quantum optics.
Classification: LCC QC793.3.A5 C53 2021 | DDC 539.7/21--dc23
LC record available at https://lccn.loc.gov/2020053103


**British Library Cataloguing-in-Publication Data**
A catalogue record for this book is available from the British Library.

For any available supplementary material, please visit
https://www.worldscientific.com/worldscibooks/10.1142/12024#t=suppl


Desk Editor: Ng Kah Fee

Typeset by Stallion Press
Email: enquiries@stallionpress.com

Printed in Singapore

This book is dedicated to my mother Betty Clarke
(20 August 1933–11 January 2020) who believed in me. She always
used to say "you're going to do well with your physics". My real
hope is that physics will be well.

# Acknowledgments

At World Scientific Publishing, I should like to thank Rochelle Kronzek for her belief in the project and Ng Kah Fee for his careful editorial attention.

The continued encouragement of Dr. John Killingbeck, University of Hull, is most appreciated. Dr. Alexander Kramida, National Institute of Standards and Technology, USA, has been exceedingly helpful in making data available, clarifying uncertainties, and providing references to further data. I am also indebted to Professor Paul Kwait, University of Illinois at Urbana-Champaign, and Dr. Mubarack Ahmed at Garden City University College, Ghana, for their correspondence on photonics.

A paper based on Chapter 3 of this book has been accepted by the peer-reviewed journal *Research Journal of Optics and Photonics*.

Friends who have expressed support for the project include Stuart Birtwistle, Norman Denton, Richard Lea, and Roger Procter.

# This World is Not Conclusion

This World is not Conclusion.
A Species stands beyond
Invisible, as Music
But positive, as Sound
It Beckons and it baffles
Philosophy, don't know
And through a Riddle, at the last
Sagacity must go ...

*Emily Dickinson*

# Prologue

This work offers an entirely new approach to the structure of the proton, the electron, and their hydrogen atom bound state. The argument I develop here is that both Maxwell's theory of electrodynamics and the present model of the hydrogen atom not only fall short conceptually, but that both are in need of a more visual basis. The classic experiment of Grangier, Roger, and Aspect (1986) is re-interpreted to suggest that the quantum mechanical wave function is based on an incorrect representation of the single photon. In fact, since there is very little that can be trusted as a secure theoretical foundation, I take as my starting point the nineteenth century experiments of Augustin-Jean Fresnel on circular polarization.

The model I develop here I denote 'mass vortex ring' (MVR) theory. It is a name that has been chosen so that the acronym points to the invariance of angular momentum in the model. However, this is not a theory based on a vortex of particles. It is actually a circularly polarized ray (denoted spin-1) passing around the surface of a toroid with two rotations: one poloidal (denoted spin-2), and the other toroidal (denoted spin-3). This is to be the form of both the electron and the proton, their difference being only one of scale. Relativistic considerations, as well as definitions of mass, charge, electric potential, and magnetic momentum arise in a natural way from this structure. Quarks, strong forces, weak forces, and gravitons play no part whatsoever in the foundations and consequences.

It seems to me that the difficulties that have hitherto obstructed progress are conceptual rather than mathematical, so the mathematics in this treatise has been kept as simple as possible. In fact, the average first year physics undergraduate fortified with a reasonable amount of persistence should be adequately equipped to follow the demonstrations. Instead, the emphasis has been placed on visual concepts with over 60 detailed figures included.

The inquisitive have been afforded the opportunity to experiment with the MVR hydrogen atom model by being given access to a working computer program MVR.exe, the BASIC code for which is given in Appendix C. This writes a detailed set of hyperfine frequency results to a file MVR.txt with a choice of 24 different hydrogen atom states. Instructions as to where the executable file can be obtained appear in Appendix C, and explanations of the program input and output are given in Appendix D.

## Achievements of the Theory

(a) The classic Grangier *et al.* (1986) experiment, which involves spontaneous parametric down conversion (SPDC), is reinterpreted to suggest that a single-photon front emerging from a beam splitter can trigger more than one detector simultaneously. This has negative consequences for the basis of quantum mechanics.

(b) An alternative theory of the Bose–Einstein counting method is given, based on rotating sequences of photons and cells in a closed cavity. It shows that the photons can be regarded as *distinguishable*.

(c) Mass is defined in terms of the perimeter of the optical orbital angular momentum (OAM) beam-waist cross-section, denoted here as a spin angular momentum (SAM) mass ring. However, this structure alone is insufficient to provide it with electrical charge, see (e).

(d) The theory of an OAM mass ring is developed. This is a toroidal structure formed by bending the optical OAM axis into a closed curve. Both the electron and proton adopt this form, the latter being a scaled-down version of the former.

(e) Electric charge sign is defined in terms of the direction of toroidal spin-3 rotation.

(f) The Coulomb law is derived from a line integral based on the relative toroidal spin-3 rotation senses of the proton and electron rings. Here, the proton spin-3 momentum field displaces energy from the electron spin-3 circuit into linear motion of the electron OAM mass ring (and vice versa).

(g) Experiments are cited showing that an electron can adopt energy levels *without* the presence of an external electric potential. This immediately calls into question the Sommerfeld and quantum mechanical treatments of the hydrogen atom. Instead, the MVR theory defines a self-potential for the electron.

(h) The MVR model is applied to the hydrogen atom to obtain an alternative derivation of the Sommerfeld fine structure spectrum, as well as *exact* agreement with experimental hyperfine frequency values in MHz (to 4 d.p.) for 24 states without recourse to quantum mechanics. There are two adjustable parameters of the model which allow:

(I) a small variation to the fine structure constant that determines the electron toroidal spin-3 speed;

(II) a modification of the eccentricity of the electron poloidal spin-2 rotation for the hyperfine calculation.

These are automatically determined by a computerized search to align the theoretical frequencies with the experimental values. Here, the Coulomb law controls the amplitude of the proton–electron bound oscillation state.

(i) The hyperfine adjustment to the hydrogen fine structure frequency arises from the changing proton spin–2 field cutting the electron spin–2 circuit at the oscillation boundary. This involves a line integral calculation.

(j) A unique feature of the MVR model is that the electric charge sign is not a property of the OAM mass ring but a behavior, and that it can be switched by altering the ring's direction of motion. Future experiments are suggested that might test this claim.

(k) Two types of acceleration are suggested: active and passive. The former is well-known, and arises from a non-absorbing and non-emitting reference frame B's observation of reference frame A's absorption of energy. However, for the latter acceleration, the absorbing frame A's observation of B will be that it accelerates *without* absorbing energy. The second type of acceleration is shown to be related to electrostatic repulsion and attraction.

Two years of my life have crept by unnoticed while developing the MVR model, and three additional years of preparatory work can be found in *The Quantum Puzzle: A Critique of Quantum Theory and Electrodynamics.* I have regularly lost valuable rest in the small hours, rising out my semi-consciousness reverie to note down promising solutions that have fermented overnight. The daily struggle has been perversely satisfying, rather like holding on to the trouser leg of a giant who has been dragging me around in the mud. Alone, I know that I can never bring this behemoth down, but my resolve never to relinquish my grip has been motivated by the hope that sharper wits might one day bring him to his knees.

There is no doubt that this has been an ambitious project, with a network of ideas so delicately interwoven that it seems inconceivable to me that it should be entirely free from error. Nevertheless, I have tested it as far as I am able, and now judge its internal consistency and predictive power to be strong enough for submission to the physics community. The MVR theory is certainly economical in construction, resting entirely on the two rotations of a circularly polarized ray confined to a toroidal surface. I sincerely believe that the inquisitive researcher will be intrigued by its new approach. At the very least, my hope is that it will suggest more productive lines of research.

In the final analysis, we physicists need to realise that we have been building houses on sand, and that there is no choice but to start all over again. The present theories of electrodynamics, quantum mechanics, atomic structure, and even dynamics carry too many unresolved issues. A new conceptual basis is needed.

Niels Bohr believed that this elusive giant could never be brought before our consciousness. I suggest that not only is it in sight, but that he must never be allowed to escape our pursuit again.

Barry R. Clarke
Oxford, UK
August 2020

# Contents

# Chapter 1

# Introduction

*Each generation of scientists gives too much credence to its own paradigms. By his education, and by participation in 'normal' science, the average research worker is heavily indoctrinated and finds great difficulty in facing the possibility that his world picture might be wrong (Ziman 1991, 90).*

## 1.1 A Grave Crisis of Ideas

How should we decide on the aim of physics? Is it merely a mathematical model for making predictions about our experience, without any visualizable mechanism? Or does it have a deeper value in trying to replicate through imaginative visualization the world external to us responsible for our experience? It is a question physics cannot answer in isolation by reference to its own partial successes or failures. If we are to find a guide as to the form our theories should take, we need to spend time examining the development of the human understanding, to discover its vector of evolution. As we shall see, as the passing millennia have provided us with an increasing sensory capacity; the human mind has produced maps of the environment that have had increasing utility. The identification and anticipation of predators, and the extraction of responses from the external world favorable to its survival, seem to have determined the direction of this development.

In his later years, one of the founders of quantum mechanics Erwin Schrödinger (1953, 52) was brought to confess that "physics stands at a grave crisis of ideas". It is a stark admission from a man who had been at the very center of twentieth century physics.

However, the difficulties facing the progress of theoretical physics did not begin with quantum mechanics but instead can be traced back to the electro-dynamical theories of the nineteenth century. George Francis FitzGerald, Oliver Lodge, and James Clerk Maxwell all attempted to construct geometrical models of the electric and magnetic fields, but all were left disappointed with their attempts.[1] Carl Anton Bjerknes subsequently saw it as an abandoned problem:

> We have theories relating to these [E-M] fields, but we have no idea whatever of what they are intrinsically, nor even the slightest idea of the path to follow in order to discover their true nature. (Bjerknes 1906, 1)

Oliver Lodge concurred:

> The problem of the constitution of the Ether, and of the way in which portions of it are modified to form the atoms or other constituent units of ordinary matter, has not yet been solved. (Lodge 1909, xix)

However, Henrik A. Lorentz had no conscience about sweeping the difficulties under the carpet, declaring that "we need by no means go far in attempting to form an image of it [the field]" and even confessed that "on account of the difficulties into which they lead us, there has been a tendency of late to avoid them altogether" (Lorentz 1916, 2).

In 1887, the Michelson–Morley experiment had failed to detect an ether wind. However, Oliver Lodge's later search for the ether was no demonstration of denial. At the turn of the twentieth century, physicists were looking for an alternative to a particle-like ether, and were trying to imagine the basic unit out of which matter might be constructed. In 1875, the English mathematician William Clifford had suggested that "matter differs from ether only in being another state or mode of motion of the same stuff" (Clifford 2011, 237).

---

[1]Maxwell adopted a model of rotating vortices, each smaller than a molecule, with intervening idler wheels, but confessed that "I do not bring it forward as a mode of connexion existing in nature" (Maxwell 1890, 486, Figure opp. 488). For details of the models by Lodge and FitzGerald, see Hunt (1991, 81, 89).

Joseph Larmor went even further suggesting that "matter may be likely a structure in the ether, but certainly ether is not a structure made of matter" (Larmor 1900, footnote vi).[2] As to the form this structure might take, Oliver Lodge suggested that electric potential energy "must be due to rotational motion [...] circulation in closed curves" and that "the speed of this internal motion [...] must be carried on with a velocity of the same order of magnitude as the velocity of light" (Lodge 1909, 102–103).

The present work develops this idea and posits a traveling screw thread (which generates circularly polarized rotation) running around the surface of a torus at the speed of light $c$. The rotation around the toroidal axis at speed $\alpha c$ is taken to be electric momentum and that around the tube axis at speed $c(1 - \alpha^2)^{1/2}$ is the magnetic momentum.[3] The magnetic field is to be in the direction of the former, while the electric field runs along the axis of the torus. This is a brief sketch of the form of a mass vortex ring (MVR) and will serve as the structure of both the proton and electron.[4]

## 1.2 Rejection of Geometrical Theories

In 1925, Werner Heisenberg invented a quantum-theoretical mechanics in which "only relationships between observable quantities occur" (1925, 168–169). His theory was based on a flawed logical positivist philosophy that gave no consideration as to how the human brain processes data, and took the view that unless quantities (such as the period of rotation of an electron in an atom) submitted themselves to direct observation, then they should have no place in a theory. Heisenberg held this to be a rigid requirement even if a theoretical model using an unobservable value correctly predicted a directly observable quantity. Schrödinger (1953, 52) thought it was "a philosophical extravagance born of despair". From that time on, the search for the geometrical structure of an underlying

---

[2]It follows from this that the idea that the quark is the ultimate building block of nature is an illusion.

[3]Here, $\alpha$ is the fine-structure constant.

[4]Perhaps it is the structure of all 'particles' but this idea will not be claimed here.

reality was completely abandoned, and Paul Dirac subsequently led physics toward a program of mathematical simplicity in theoretical constructs with the declaration that "the main object of science is not the provision of pictures but it is the formulation of laws governing phenomena" (Dirac 2000, 10).

As we shall see below in our discourse on the evolution of human understanding, we need to be clear as to the aim of physics. If it is the mere description of phenomena, then, since it contains unconscious prejudices connected with the way our human sensory apparatus works, it is actually the most primitive form of theorizing possible. However, if it is an attempt to mirror the unobservable structures causing the phenomena, then the adherence to sensory concepts can yield no progress. Of course, there can be no direct comparison between our conjectured model and the external structure it purports to represent. The model's justification rests instead on its inner consistency and its success in producing consequences that correspond to observed phenomena. As our visual model ascends to greater success, we can entertain increasing confidence that this is the form of the world beyond our senses causing our experience.

Before discussing the type of theory that should be pursued for the best chance of theoretical success, there are a number of basic problems in physics that are far from having a satisfactory solution. For example, recent experiments on electron vortices suggest that an electron can adopt energy levels even in the absence of an external potential (Uchida and Tonomura 2010; Verbeeck *et al.* 2010). McMorran *et al.* (2011, 194) have concluded that "electrons can be prepared in quantized orbital states with large OAM, in free space devoid of any central potential, or medium that confines the orbits". However, if the external potential function arising from the proton is removed from the Hamiltonian function in the quantum mechanical treatment of the hydrogen atom, it is completely deprived of its energy levels. The new MVR model presented here seeks to remedy this by positing a self-potential for the OAM mass ring.

As far as the theory of electrodynamics is concerned, it is surprising that confidence is still placed in a theory that was conceived over 30 years before evidence for the electron was first secured (Thomson 1897),[5] and over 70 years before light rays were demonstrated to possess spin angular momentum (SAM) (Beth 1936).[6] Maxwell's equations are ill-equipped to accommodate either of these concepts. In addition, there is a lack of explanation in these equations. For example, all we have in Ampère's law is a law of association between a current and a surrounding magnetic field. It makes no attempt to describe a mechanism for the phenomenon. All it claims is that when a current is activated, a magnetic force circulates around it and when it is deactivated it vanishes.

On the basis of this law of association, Richard Feynman has given an argument as to why a mechanism cannot be found for the magnetic field associated with a moving charge (Feynman *et al.* 2006, II.1–10). Let us first set out two premises as follows:

(a) A moving charge *creates* a magnetic field in virtue of its motion, and it is a field that has no existence when the charge is stationary.
(b) The magnetic field has an underlying mechanism.

Feynman gives the example of two identical free charges moving parallel to each other at the same speed. He considers an observer A stationary in some reference frame who sees the two charges moving and therefore observes that each charge is surrounded by a magnetic field. A second observer B who is moving with the charges views them both as stationary and so detects no magnetic field.

Feynman argues that given the truth of (a), then (b) can be true for A but not for B. If the existence or non-existence of a mechanism depends on the state of motion of a charge, then this is unrealistic. So, Feynman rejects (b) for *all* observers while retaining (a).

---

[5]Thomson measured the charge-to-mass ratio of cathode rays in an evacuated tube.
[6]Beth used a suspended half-wave plate to reverse the rotation sense of circularly polarized light and create a measurable rotational reaction in the plate.

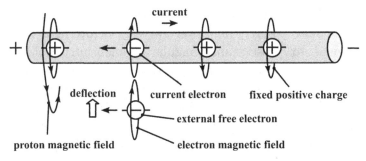

Figure 1.1  The MVR theory explanation of a magnetic field. A potential difference across the wire ends aligns the proton and electron ring axes and their oppositely rotating spin-3 magnetic fields. The electron is deflected toward the wire by a Lorentz force.

Now Ampère's law refers to charges in a wire, so let us place Feynman's argument in this context before returning to his example; see Figure 1.1. Let one of the two electrons be part of the current in a wire and let the other electron be free and external to the wire yet moving parallel and at the same speed as the first one. Allow observer A to be stationary with respect to the wire, while B is moving with the two electrons.

Now, according to the MVR model presented here, we intend to suggest that (b) is true but (a) is false, and that (a) should instead be modified to the following:

(a′)  A charge has a magnetic field surrounding it both at rest and in a state of motion.

In fact, this will be a property of our OAM mass ring. In that case, one is now permitted to posit a magnetic field mechanism for all observers. With this in view, we shall proceed to explain the deflection of the external electron toward the wire as a Lorentz force effect.[7] It is a deflection that must exist for both A and B.

However, first we must ask the following: Why does the magnetic field vanish when there is no current? In a wire with no potential

---

[7] The magnetic vortex field and the Lorentz force effect in Ampère's law are developed in greater detail in Clarke (2017, §8.3).

difference across the ends, the charges are stationary yet the novel suggestion made here is that magnetic fields permanently exist around both the protons and electrons. They do so as part of the proton and electron ring structure, not as a consequence of their motion. For a copper atom in the wire, not only is the magnetic field rotation of an electron equal and opposite to that of a proton but the atoms in the wire are randomly orientated so that the fields from different lattice sites cancel out. So, there is no net magnetic field around a copper atom, and a wire with no moving electrons has no *detectable* surrounding magnetic field.

Now, when a potential difference is applied to the ends of the wire, the axes of the magnetic field rotations line up along the wire. The axes of the moving electrons become aligned in parallel, while the axes of the stationary positive charge sites become aligned anti-parallel to them. The randomness of orientation is no longer responsible for the cancellation, but so long as nothing moves, there is still a cancellation of equal and opposite rotations, and there is no relative motion of oppositely rotating magnetic fields. When the current flows a moment after alignment, the electrons leave the positively charged sites and the aligned positive-charge magnetic fields surrounding these sites are exposed. We now invoke the Lorentz force, see Figure 1.1. For observer A, the free external electron (traveling parallel with the current electron in the wire) moves through the magnetic field of the protons that runs perpendicularly to their aligned axes (but not through the relatively stationary electron magnetic fields from the wire) and is thereby deflected toward the wire. From the point of view of observer B traveling with the free external electron, the wire moves and the proton magnetic flux lines pass through the electron (while the electron fields from the wire remain relatively stationary) and so the electron is still deflected the same way. In other words, the force arises from the relative motion between an electron charge and a proton magnetic field and it is a relative motion that exists for both observers A and B.

Let us now return to Feynman's example. If we remove the wire, and set two free electric charges in parallel motion at the same speed, for both A and B, although they each have a structural magnetic field surrounding them, there is no *relative motion* between a charge and a magnetic field, and so there is no Lorentz-type force between them. For this reason, we should not expect the electrons in cathode rays to interact magnetically with each other. We shall see in Chapter 6 that an electrostatic interaction between MVRs requires that the ring axis of one penetrates the ring of the other. On that basis, there is no electrostatic interaction either between Feynman's two parallel moving electrons.

## 1.3    Evolution of Human Understanding

Let us now focus on the way human knowledge of its environment has evolved. The theory as to how the brain attains knowledge has traditionally been the province of epistemology where only internal data have been available to the philosopher in the form of what he perceives, thinks, and feels. The arguments in Immanuel Kant's *Kritik der reinen Vernunft* (*Critique of Pure Reason*, 1781) were developed entirely on that basis. However, epistemology is now a growing science and if we are to make progress in our task of approaching a complete understanding of nature, account should be taken of the results of experiments on our human processing apparatus. This lends important guidance as to the form our theories must take in order to succeed in our task of bringing sense impressions into the greatest order, and making reliable and penetrating predictions about Nature.

The manner in which humans develop knowledge is the evolutionary product of the interaction between the information-processing subject and the external world beyond the senses. Information is evaluated in relation to the survival of the subject, or more fundamentally, the elusion of pain. For example, for the avoidance of predators and for the selection of nutritious food, those creatures who have built mental models that best correspond to the external world have endured. As Konrad Lorenz states,

> The 'spectacles' of our modes of thought and perception, such as causality, substance, quality, time, and place, are functions of a neurosensory organization that has evolved in the service of survival. (Lorenz 1978, 7)

The following question arises: What is the process by which knowledge of this external world is secured? Let us explore this problem a little further before returning to our main project, the construction of a workable atomic model for hydrogen.

The processing subject has a predisposition to identify invariant features in experience when it is faced with contingent events. In order to make decisions, it needs information it can depend on. For example, the form of a chair is still recognizable whatever angle it is seen from. It is not seen as several different chairs. The color of an object is perceived as constant despite varying lighting conditions. The size of an object is unchanging even as it recedes from view. These are perceptual invariances.

On a deeper level, the laws of physics are held to be the same for all systems in uniform motion. It is the absence of variation and contingency and the presence of a common pattern or a persisting characteristic that mark out the objectively real. Feelings and wishes that have been associated with previous experiences have been stored in memory, and are imposed on similar new ones in order to speed up the decision making as to what should be approached or avoided. These evaluations intrude on our objective assessment. The more we can separate out these subjective contributions, and distance ourselves from the egocentric, the closer we can get to an image of external reality which exists independently of us. It is a process of extracting the subjective manner in which we process data.

Even at the level of visual perception, we can see that we are not merely passive receivers of facts. A conjecture is *imposed* on ambiguous data as a tentative best fit (postulate and test) rather than information being *extracted* from given sense data (induction). For example, it appears that there are already rules in place for interpreting groups of images before they arrive. Richard Gregory has developed the idea that examples of visual illusions support the presence of a 'hypothesis generator' for visual perception "to compensate neural signalling delay [...] so 'reaction time' is generally avoided" (Gregory 1997, 1121). Here, previous perceptual knowledge in memory is compared with new images to obtain the best possible interpretation. He cites the idea of an upright rotating face mask that alternates to our view between convex and concave, Figure 1.2.

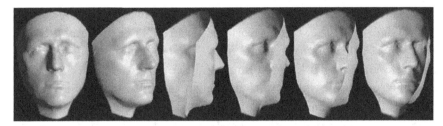

Figure 1.2    The rotating face mask. The first image is of the face convex as expected, and as the front of the mask rotates to the right we reach the last image which is concave. However, it is still understood as a convex face.

Through previous experience, we associate the idea of convexity with a real face. However, even when the angle of rotation is such that the view is concave, the mask still looks to us like a real face and so we falsely understand it to be convex. So Gregory suggests that there is a hypothesis generator acting on perception that is imposing the interpretation 'this face is convex' onto the concave image.[8]

> This bias of seeing faces as convex is so strong it counters competing monocular depth cues, such as shading and shadows, and also very considerable unambiguous information from the two eyes signalling stereoscopically that the object is hollow. (Gregory 1997)

Examples of visual illusions, and our errors in classifying them, are evidence of this hypothesis generator in action.

Although Immanuel Kant in the *Kritik der reinen Vernunft* (*Critique of Pure Reason*, 1781) decided that the external world or 'thing-in-itself' existed, he concluded that nothing could be known about it. This was his 'transcendental idealism'. For Kant, only direct experience was knowable and this arose from the application of our processing apparatus to given sense data. Unfortunately, there is no room for conjecture in Kant's theory, and so there is no opportunity to rise to more abstract and more widely applicable theoretical schemes. However, he appears to have been a pioneer in developing the interpretive character of human knowledge.

---

[8] "[...] perceptions are regarded as similar to predictive hypotheses of science, but are psychologically projected into external space and accepted as our most immediate reality." (Gregory 1997)

From birth, our apparatus is adapted to process our sense data, that is, even before receiving any input. In this sense, we possess *a priori* knowledge. It applies boundaries to the light-intensity map on the retina, groups together a set of contiguous moving images into a time-sequenced object, and then imposes causal laws on objects coincident in space and time. However, our sensory-processing equipment has acquired its form over a long period of selective adaptation in relation to the objects in the external world, and those human processors that have constructed the best models of this world have had the greatest success in extracting desirable consequences from it.

In this sense, we can know something about the thing-in-itself because our success in surviving is evidence that the models that we have adopted are a good approximation to the unobservable external world. The better our ability to understand the operation of the world independent of us, the better will be our predictions as to how it will behave in certain circumstances, and the better will be our chances of survival. The evolution of our modeling can be approximated by observing lower species. As Konrad Lorenz states, "it is possible to make statements as to whether agreement between appearance and actuality is more exact or less exact in comparing one human being to another, or one living organism to another" (1982, 235). Those lower on the evolutionary scale should exhibit less agreement.

There is the logical positivist view that only what can be observed is 'real'. However, I still believe my house exists when I am away from it, even though I am not continuously observing it. This means my belief is only a conjecture which I retain only because of its utility. The conjectured invariance of my house's existence is something that I regard as real, as it would be difficult to function without this confidence. However, being conjectural in character, my belief is always subject to the critical test of my observing my house when I return. So long as it survives this repeated test, then I retain confidence in the belief.[9] However, if I returned home one day and found that my house had gone, then I would be forced to modify it.

---

[9]Should I live in a war zone where regular bombing occurs, I might not hold this belief with the same degree of confidence.

There is also a philosophical creed known as 'naïve realism' in which it is believed that the world external to me exists in the same form that I perceive it. It holds that we simply receive a copy of it through our senses. The following considerations should illustrate why this view cannot be correct, although for most of our everyday interactions with the world, it is sufficient to assume that our perceptual images are mere copies of the external world.

## 1.4   A World of Changes

In 1885, Balmer (1885, 83) published a formula that corresponds reasonably well with the known experimental measurements for hydrogen lines in the visible and ultraviolet region

$$H = \frac{m^2}{m^2 - 2^2} h \tag{1.1}$$

where $H$ is the wavelength, the integer $3 \leq m \leq 11$, and $h(10^{-10}\,m)$ is a constant. It was generalized three years later by Rydberg (1889; 1890, 333), for all hydrogen lines, and was presented as a wavenumber

$$n = n_o - \frac{N_o}{(m + \mu)^2} \tag{1.2}$$

where $m$ is any positive integer, $N_o$ is a constant common to all series and elements, and $n_o$ and $\mu$ are constants particular to the series.[10] The calculation of a spectral line results from the difference between two different values of $m$ in (1.2). It should be clear from this that the light we receive into our sensory apparatus results from the difference between an initial and a final state. The object, or 'thing in itself' as Kant called it, is not given to us directly. We do not receive a copy of it through our sensory apparatus. Only the *changes* in that object are presented to us, and these changes take the form of light signals. It is rather like observing a man depositing and withdrawing money from his bank account in predictable amounts without ever getting to see the balance. If one wants to know the balance, then

---

[10] Balmer's formula results from the choice $\mu = 0$, $N_o = 4n_o$, and $n_o = 1/h$. Then, $H = 1/n$. $N_o$ became known as the Rydberg constant.

a model of the account needs to be imagined that will predict the given transactions.

## 1.5 Evolution of Vision

So, let us start at the most primitive part of our human understanding, that which is directly related to our visual apparatus. An organism develops senses to locate obstructions and threats in its environment. Those that possess the best detectors have the best chance of survival. A large part of our understanding of the external world has arisen out of the evolution of the eye and we now take examples from more primitive life forms to illustrate it.

A survey of the evolution of visual receptors can show how maps of the external world have achieved increasing precision. Eye type can be classified into four cases: (1) non-directional photoreception; (2) directional photoreception; (3) low-resolution vision; and (4) high-resolution vision (Nilsson 2013, 10). We now examine each case.

(a) Cyanobacteria possess unscreened photoreceptor pigments and have been found to exhibit phototaxis[11] according to absorbed light intensity (Häder 1987, 1, 12). At low intensities, they move toward the light source and at high intensities they move away. Their ability to migrate according to lighting conditions appears to optimize their ability to conduct self-preserving photosynthesis. Even with this primitive information, the cyanobacteria can still 'know' something about the unobservable source emitting the light, in this case its approximate direction.

(b) The pear-shaped larvae of a box jellyfish *Tripedalia cystophora* have 10–15 ocelli. These are cup-shaped structures that are evenly spaced on their posterior and are filled with photosensitive pigment. The cups screen light from certain directions and form a directional light meter. As the larvae rotate at 2 revolutions per second about their longitudinal axis, they continuously scan their

---

[11]Phototaxis describes a direction of motion with respect to the direction of the light source. Photokinesis describes the speed in relation to the light intensity.

environment to obtain a rough map of the spatial distribution of light (Nordström *et al.* 2003, 2351). So, the directions of several simultaneous light sources now become knowable.

(c) Low-resolution eyes usually have a lens at the entrance to a cup-shaped depression, but the retina is too close to it to yield a high-resolution image. Its original function might have been to "prevent foreign material to enter the eyecup" (Nilsson 2013, 12). An example is *Chiropsella bronzie*, the box jellyfish, which has four sensory indentations, or rhopalia, that between them carry 12 pigment pit eyes, 12 slit eyes, as well as four upper and four larger lower lens-eyes. Let us consider only the eyes with lenses. The lower lens-eye has an iris with a focal length that falls well behind the retina. With a spatial resolution of no more than 20°, it can only detect "very large structures at close range" (O'Connor 2009, 563). So, in addition to directional information, the box jellyfish now has access to simple close-distance data that it can incorporate into a primitive map of objects in its environment.

(d) Eyes capable of producing high-resolution images are given to vertebrates, cephalopods, and arthropods. This improvement allows the type of predator to be distinguished so that a judgment can be made on its threat-status and an appropriate response can be enacted. As the rate of data delivered to the nervous system increases, and the light-intensity map on the retina possesses greater detail, there is the need for an increase in brain-processing power.

## 1.6  'Particle' is a Sensory Prejudice

So far, we have examined the most primitive form of interpretive understanding, that is, the direction, distance, and identity of light-emitting sources. This discrimination is clearly connected with the survival of the host. The next stage involves the processing of a high-definition visual map of the environment. A human brain is an electro-chemical information-processing unit confined to a bounded volume inside the skull. If we are to assess the justification for many of the concepts that enter our physical theories, we need to pay attention to how they have arisen.

How do we get from a 2D array of image-intensity values provided by the retinal photoreceptors to our experience of 3D bounded objects complete with an anticipation of the complex interactive relationships between them? The inputs to the brain are supplied by visual, aural, olfactory, gustatory, and tactile sensors. As far as visual sensation is concerned, the two eye lenses produce a highly focused image, one on each retina (Delbrück 1986, 95–108).[12] The slight displacement of these two images allows a reasonable computation of the direction and distance of the center of the light source.

The retina has about $10^8$ photoreceptor cells; see Figure 1.3. A small area of these light-stimulated cells trigger a bipolar cell, several of which supply a retinal ganglion (RG) cell that surveys a circular area of about 100 photoreceptor cells. This circular area is divided into an annulus and an inner circle, and an RG cell is designed to compute the light contrast between the two regions.

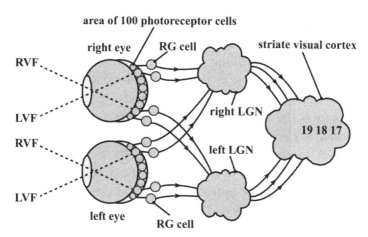

Figure 1.3   Plan view of the human brain showing the right visual field (RVF), left visual field (LVF), retinal ganglion (RG) cell, right and left lateral geniculate nucleus (LGN), and the striate visual cortex.

---

[12]Delbrück provides a clear account of our processing apparatus.

The $10^6$ RG cells separate into two types: one type generates a strong signal when the annulus receives greater illumination than the inner circle, and the other when the inner circle is brighter than its surrounding annulus. As we shall see, the function of these cells is to supply clear boundaries to patches of light on the retina.

The RG cells in the left part of the visual field from both eyes transmit to the right side of the brain, to a region known as the right lateral geniculate nucleus (LGN). Similarly, the cells in the right visual field send signals to a similar area in the left LGN. These converge on simple and complex cortical cells, each one being supplied by a set of RG cells. For example, one such set might be constructed from RG cells that are supplied by photoreceptor cells lying in a straight line that has a particular orientation in the visual field. In fact, each cortical cell is dedicated to a particular line orientation or direction of motion of an orientation.

Area 17 in the striate visual cortex at the rear of the brain is dedicated to pattern recognition, and there we have the construction of continuous lines, linear or curved, indicated by areas of contrast. Information is then sent to regions 18 (visual area V2) and 19 (visual area V3) further forward where a primal sketch of the visual scene is created (Marr 1982, Table 1-1). From this, a so-called $2^1/_2$D sketch is assembled containing information about the distance from the viewer, discontinuities in depth, and discontinuities in surface orientation. Finally, a 3D model is produced.

Now, here is the crucial point. Our sensory-processing apparatus has imposed a clearly defined boundary around an 'object' that did not exist in the original retinal intensity distribution.[13] Since the visual map presented to our consciousness arises from the light emissions from the source object, there is no necessity for the source object itself to be bounded. Recall that we only receive *changes* in the external object not the object itself. It might possibly have unlimited extent. Now, at first glance, our tactile experience appears to reinforce the notion that an object is bounded. We reach out to

---

[13]Here, the 'object' is defined as an area of light intensity on the photoreceptor mosaic that is brighter than its surroundings.

touch the desk and the motion of our finger is arrested at the visual boundary of the desk. However, this 'boundary' is only the region where the electrostatic force of repulsion between our finger and the object is sufficient to halt our motion toward it. It need not be a boundary-to-boundary contact as our visual experience leads us to understand.

Nevertheless, it is precisely this tactile resistance that fortifies our belief in an external world. If we were only conscious of a 3D visual map, we might be tempted to conclude that we are passive observers of a mere hallucination. We might then believe that an external world spatially separate from us is only our mental creation. However, not only would this make our possession of sensory apparatus superfluous but we would also need to account for the fact that before the age of 5–6 months, the human child intentionally reaches out toward a visually bounded object and finds resistance to the touch in the vicinity of the visualized boundary (Piaget 1973, 54–55, 66–67).[14] It is this coordination (or apparent mutual corroboration) of visual and tactile space boundaries, supplemented by directed auditory stimuli, that gives the child confidence in positing a world that exists in a different part of space to itself. Eventually, it is noticed that events occur that are both contrary to his will and contingent. There is a world that exists independently of his processing apparatus. This suggests that Kant's 'thing in itself', the source of the light signals, is *really* rather than *apparently* independent of us.

Is it possible that everything I 'know' is a product of my own imagination? There is an argument against this solipsist position. If there were not an external world spatially separated from my brain, then although my mental apparatus presents the works of Mozart, Newton, and Shakespeare to me as not being mine, my brain would still have to be capable of generating them by itself in all their intricate detail. However, in principle, when future medical technology permits, a surgical exploration of my brain could be carried out which could demonstrate its incapacity to perform such feats of originality. This fact would then lead to an inconsistency.

---

[14]Piaget calls this the "coordination of vision and prehension".

One could suggest that this medical examination might itself be part of the hallucination, and so brings no authority to the judgment. However, if my goal is to produce a logically consistent explanation for my experience, I would have to reject my solipsist position. I would then have to conclude that these works are not created by my brain but are presented to it by a source external to it. One would then have to decide if this source were supernatural, which has the advantage of needing no further thought, or material, in which case considerable effort would be needed to imagine its structure.

The idea that one is not the entire world and that there are objects that exist independently of oneself is an early lesson in human development. Before two years of age, the child comes to believe that the world conforms to his will, a view that is initially reinforced by an attentive and obliging mother. However, he is soon exposed to contingent events that are beyond his control. Piaget (1971, 397) remarks that in the second year of life, "The self [...] discovers itself as a cause among other causes and as an object subject to the same laws as other objects".

It should be clear by now that experience contains elements of human sensory processing that must be given up if we are to get closer to a model of the 'thing-in-itself', the unobservable world that gives rise to our retinal light-intensity distribution. As we abandon sensory concepts such as a bounded or particle-like object, as well as the survival-orientated feelings attached to them, the higher-level and more widely embracing ideas acquire a greater invariance.[15] The need to abandon egocentrism was clearly expressed by Konrad Lorenz:

> Every time we succeed in tracing an element in our experience to 'subjective' factors, and in then excluding it from the image we form of extra-subjective reality, we come a step closer to that which exists independently of our cognition. (Lorenz 1978, 3)

---

[15]Max Planck said, "This constancy, independent of all human and intellectual individuality, is plainly what we call the real" (quoted by Ryckman 2017, p. 331). Survival-orientated feeling can also obstruct the adoption of new and more promising ideas.

## 1.7 A Return to Visualizable Concepts

Although symbolic mathematics is the language in which this 'image we form of extra-subjective reality' is to be conveyed, being an atomic mechanism, its expression can only take a geometrical form. Both Bohr (1913) with his circular-orbit model of hydrogen and Sommerfeld (1916) with his elliptic-orbit model realized this and both posited a visualizable atomic structure. They were notable approximations to reality because whatever these structures absorbed or emitted and whatever their state of motion, they always retained the same form. However, Bohr was confounded by the wave-particle duality, and in an act of resignation contrived to save his own sanity with the assertion "evidence obtained under different experimental conditions cannot be comprehended within a single picture" (Bohr 1951, 210). So, he supported the Heisenberg position in which only concepts relating directly to observables are permissible.[16] This emphatically prohibits the construction of visualizable models, and paralyzes the only method of devising closer and closer approximations to the thing-in-itself.

Unfortunately, Bohr's personal authority on this matter has impeded the progress of theoretical physics for the last 100 years. Even 400 years ago, Francis Bacon knew of such men:

> they prefer to blame the common condition of man and nature rather than admit their own incapacity (2000, 8) [...] they turn the weakness of their own discoveries into an insult against nature itself and a note of non-confidence in other men (2000, LXXV, 62) [...] this is wholly due to a wilful limiting of human power, and to an artificially manufactured desperation, which not only dims any visions of hope, but also blights all the incentives and nerves of industry [...] (2000, LXXXVIII, 73)

---

[16]Heisenberg held the view that any quantities that were unobservable had no place in a theory and claimed that "it appears better to give up the hope of an observation of the previously unobservable quantities (such as the position and time of revolution of the electron [in a hydrogen atom]) [... and instead] to attempt to formulate a quantum-theoretical mechanics analogous to classical mechanics in which only relations between observable quantities occur." (Heisenberg 1968, 168–169).

For Heinrich Hertz, physical theories might well contain an image designed to approximate the thing-in-itself, but the former cannot be directly compared against the latter to discover how close to 'reality' it is. Instead, he gave criteria for deciding between models, especially ones that predict the same phenomena:

> The images [in models] which we may form of things are not determined without ambiguity by the requirement that the consequents of the images [in the theoretical model] must be the images of the consequents [representations of phenomena]. [...] We should at once denote as inadmissible all images which implicitly contradict the laws of thought. [...] We shall denote as incorrect any permissible images, if their essential relations contradict the relations of eternal things. [...] Of two images of the same object, that is the more appropriate which pictures more of the essential relations of the object [in experience ... and contains] the smaller number of superfluous or empty relations — the simpler of the two. (Hertz 1899, 2)

He realized that these images are not extracted in any way from experience since "our requirement of simplicity does not apply to nature, but to the images thereof which we fashion" (Hertz, 1899, 24). However, despite stating the logical independence of the theoretical model and the images in experience,

> Hertz renounced the theoretician's supposed need for heuristic resemblances or visualizable models and promoted a new abstract conception of a physical theory in which the sole relation between the premises of the theory and entities or processes in nature need only be symbolic. (Ryckman 2017, 325)

Influenced by Hertz, Einstein set out his program of theoretical physics as follows:

> The essential thing is the aim to represent the multitude of concepts and propositions, close to experience, as propositions, logically deduced from a basis, as narrow as possible, of fundamental concepts and fundamental relations which themselves can be chosen freely (axioms). [...] Physics constitutes a logical system of thought which is in a state of evolution, whose basis cannot be distilled, as it were, from experience by an inductive [logical] method,

but can only be arrived at by free invention. [...] The justification (truth content) of the system rests in the verification of the derived propositions by sense experiences [... and] evolution is proceeding in the direction of increasing simplicity of the logical basis. (Einstein 1954, 294, 322)

As Ryckman observes (2017, 350–351),

the ideas of theoretical reason [...] give expression to reason's capacity to surpass the confines of experience through the hypothetical adoption of maxims of systematic unity or unity of nature [... which ...] any current state of knowledge of the world only approximates.[17]

Two main points arise out of the foregoing considerations. First, Newton's classical program of recording the position and momentum of every object in the local universe to predict its future evolution through a system of mechanical equations cannot be carried out. This should have been evident long before the invention of quantum mechanics, when the difference between terms in spectral line formulae suggested that it is not the thing-in-itself that is given to the observer but changes in it. For this reason, obtaining the precise location of the emitter of these changes from the changes themselves is impossible. Second, if we subscribe to the Heisenberg–Bohr philosophy of confining ourselves *only* to concepts that directly relate to observables, then we are committed to the 'particle' representation of light. However, this is a prejudice that arises out of the way our sensory processing represents objects as bounded volumes. In that case, there is no theoretical freedom to posit an alternative structure of light.

The instrumentalist doctrine that one should not seek causal explanations in terms of models, but only descriptions of experience, appears to have originated with Pierre Duhemat at the start of the First World War. According to Duhem, "physical theory is not an explanation, but a simplified and orderly representation

---

[17]For empirical science to exist, the assumption is needed that reason's striving for simplicity and unity corresponds with a world independent of the mind.

grouping laws according to classification which grows more and more complete" (Duhem 1954, 54). However, this is rather like finding similarities in the blocks that form only the foundations of a pyramid. As we have seen, if we are to make progress in obtaining knowledge about the world beyond the senses, the human mind must build this pyramid into increasing levels of visualizable abstraction from the senses, heights from which the lowest levels of the structure supporting it can be predicted.

## 1.8   The Nature of the Photon

We have already seen the argument that the notion of mass as a bounded and filled volume is a prejudice arising from the way we process sense data.[18] These sense data are presented to us as a change in the object external to us from which they originated. So, if we wish to know the structure of the originating object, we must be prepared to abandon concepts that are only useful for the most primitive level of our understanding. This requires a construction which finds its justification not in its direct comparison with the external object — for that is impossible — but in how well it can predict the data we receive through our senses. The recommendation here is that this construction should be geometrical in form, a basic atomic mechanism capable of producing the observed line spectra. As Popper observes, this is exactly what Bohr (1913) attempted but later abandoned when he realized he could make no further improvements:

> [...] nobody could have been more keenly aware of the depth of the difficulties that beset his [Bohr's] atomic model of 1913. He never got rid of these difficulties. When he accepted quantum mechanics as the end of the road, it was partly in despair. (Popper 2000, 9)

The structure of mass and charge is still an unsolved problem. Lorentz thought the electron was a solid sphere where the charge was "distributed over a certain space, say over the whole volume

---

[18]We are confined to thinking visually in space and time, there is no way out of this, but the challenge is as follows: Can we produce a structure in this form from which all phenomena can be deduced?

occupied by the electron" (Lorentz 1916, 11). He later modified his view, suggesting that "the charge might be distributed over a thin layer on its surface" (Lorentz 1916, 16). What this smeared out charge consisted of he had no conception. Feynman also viewed the electron as a sphere stating that "the field from a single charge is spherically symmetric" (Feynman 2006, 1–5). However, since the charge — the smeared out unidentified agent — can be divided into parts which react with each other, then this kind of electron must have a self-energy that arises from bringing these parts together from an arbitrarily large distance. As Fermi recognized, "the problems connected with the internal properties of the electron are still very far from solution" (Fermi, 1932). This is still true today.

So, where do we start in constructing our model of the electron? In the treatise that follows, we begin with circularly polarized light, for if we wish to understand the most recent experiments on the spin angular momentum (SAM) and the orbital angular momentum (OAM) of light, then it is indispensable to optics; see Chapter 5. In Chapter 2, the theory of circularly polarized light is developed and certain quantities such as SAM density, linear momentum density, and energy density are derived from it and compared with Maxwell's theory. Once we have this as a conceptual foundation, our next problem is to isolate a single photon and analyze experiments on photonics to see what adjustments need to be made to the idea. This is the aim of Chapter 3.

Experiments using spontaneous parametric down conversion (SPDC) generate an idler and signal photon pair. This is the set-up of the Grangier *et al.* (1986) experiment, and it is reinterpreted to suggest that a single photon consists of an advancing array of parallel circularly polarized tubes, each capable of exciting a detector. There are two probabilities to consider here. The first is the probability that the signal photon excites a detector given that it is incident on it, known from the detection of its idler twin at another detector. From the experiments analyzed, this probability is about $7.5 \times 10^{-4} - 1.0 \times 10^{-2}$. The second concerns the separation of the signal photon at a beam splitter into two detectors. We are then interested in the coincidence probability that one detector records a hit given that the other has registered. This is about $5.0 \times 10^{-5} - 3.5 \times 10^{-4}$. The second registration is usually interpreted to be an intruding photon from an unrelated SPDC event, but this view will not be adopted here. Here, the coincidence will be taken as evidence that a single photon front is capable of multiple detector registrations, an occurrence that has such low probability that it has hitherto passed unnoticed. The photon front will be seen as an array of advancing screw threads or helical space dislocations (HSDs) transversely iterated.

Having covered the transverse iteration of a circularly polarized tube, in Chapter 4, we treat the longitudinal iteration. This is the idea that $n$ tubes can be joined end to end with action $nh$. The Bose–Einstein counting result is given a novel reworking to show that the idea of indistinguishable elements is not necessary for its derivation but it can be produced with distinguishable photons. This involves a rotating sequence of photons and cells in a closed cavity in which a cell is defined by the direction that a photon approaches it. It is shown that in a cavity containing a known sequence of photons with distinct frequencies in which the photons follow a closed path, if one photon can be identified in one cell, then the possible distributions of the remaining photons among the remaining cells can be ascertained based on the assumption of distinguishability.

Chapter 5 covers the theory of optical OAM. A transverse array of circularly polarized tubes (HSD) can be modified by optical elements to produce light with OAM. This creates what we shall call here a 'helical array dislocation' (HAD) which amounts to a rotation of HSD around the optic axis. Here, the Poynting vector possesses both a linear and azimuthal momentum, the latter momentum being a feature that Maxwell's theory of electrodynamics is at a loss to accommodate.

## 1.9  The Mass Vortex Ring

Chapters 6–9 develop the theory of the MVR. To some extent, this was carried out in *The Quantum Puzzle* (Clarke, 2017) but only for an 'unloaded' ring, that is, one that has absorbed no radiation. The focus in that work was on electrodynamics where a derivation was provided for the Lorentz force using a changing parallel proton spin-2 momentum field, and explanations were given for the deflection of currents in parallel conductors as well as electromagnetic induction.

The first step in constructing an OAM mass ring is to take optical OAM, which is to be called spin-2 here, and define mass in relation to its minimum beam waist. This is to be called an SAM mass ring. It is assumed that the space surrounding this minimum radius also contains angular momentum, where HSD tubes are stretched out to provide field momentum. The axis of the SAM ring is then bent round into a closed circle so that the SAM ring runs along the surface of a torus or OAM mass ring. This rotation is to be called spin-3 and its existence introduces charge to the mass. Electrons and protons (and presumably other 'particles') are constructed in this manner, the proton being a scaled-down version of the electron with an opposite spin-3 rotation sense. Coulomb's law is derived on the basis of these OAM rings. Also, a modification to dynamics is proposed in which it is suggested that there are two types of acceleration: active and passive. The first is well known (though not by this name) in which a mass ring $A$ absorbs energy in order to accelerate, and is observed to do so from a non-absorbing mass ring $B$. The second type is observed from the absorbing system $A$, and here $B$ accelerates without absorbing energy, as a consequence of $A$'s absorption. The relationship of $B$ to electrostatic repulsion and attraction is pointed out in Chapter 6.

In Chapter 7, our attention is directed to the absorption of radiation to create a 'loaded' OAM mass ring. At this point, a defect in both the Sommerfeld and quantum mechanical treatments of the hydrogen atom is pointed out. They both depend on an external potential to create hydrogen energy levels, yet, as stated earlier, there are experiments that show that an electron can have energy levels in the absence of such a potential (McMorran *et al.*, 2011). So, the Sommerfeld derivation of the fine-structure formula is modified here to exclude the external potential and include a self-potential. It turns out that if half of the OAM ring energy is radiated at the boundary to the hydrogen atom state, the fine-structure spectral line formula can be recovered to a slightly better accuracy than before.

The proton–electron bound state is developed further in Chapter 8 where certain principles are set out for its operation as follows.

(i) All emissions and absorptions occur at the maximum oscillation amplitude of the electron.

(ii) For a transition to occur, an excited OAM ring must temporarily collapse to the ground state circular radius.

(iii) All emissions are subject to a Doppler effect based on the electron's ground state ring in which the proton's speed of approach is taken into account. This can either be a red shift, in which the radiation emission passes through the proton before exiting, or a blue shift in which the emission moves directly away from the proton.

(iv) The proton and electron oscillations are $\pi/2$ out of phase.

To execute an accurate fine-structure calculation, a *Kalpha* parameter (which multiplies the fine-structure constant) is varied to bring the state frequency into alignment with the average of the two experimental hyperfine values. Finally, Chapter 8 suggests how electron configuration and alpha particles might be fitted into the MVR theory.

The final step in the MVR treatment of the hydrogen atom is to raise or lower the above-mentioned average hyperfine frequency to obtain the exact high or low hyperfine frequency for the state in view. This is the aim of Chapter 9. Here, the rate of increase of the proton's spin-2 momentum field is taken into account, and the eccentricity of the electron's spin-2 circuit is varied for precise results. Same-sense proton and electron spin-2 rotations (input $+$) produce the low hyperfine value, while opposite rotations (input $-$) give the high value.

So, exact hyperfine frequencies essentially depend on two parameters: *Kalpha* which varies the fine-structure constant, and the electron spin-2 eccentricity which varies the strength of proton field momentum absorbed by the electron. A computer program is provided for the reader which automatically searches for these two parameters. There are 24 states available for input, as well as the two relative proton–electron spin-2 rotation senses (+ for the low hyperfine, − for the higher), the angle that the electron spin-2 major axis is orientated (usually 135° suffices), and the type of Doppler shift (red or blue).

## References

Bacon, F. *The New Organon (1620)*. Cambridge University Press, 2000.

Balmer, J. J. 'Notizüber die Spectrallinien des Wasserstoffs [Note on the spectral lines of hydrogen]'. *Annalen der Physik and Chemie*, Band 25 (1885): 80–87.

Beth, R. A. 'Mechanical detection and measurement of the angular momentum and light'. *Physical Review*, 50 (1936): 115–25.

Bjerknes, V. F. K. (ed.). *Fields of Force: A Course of Lectures in Mathematical Physics Delivered December 1 to 23, 1905*. New York: The Columbia University Press, 1906.

Bohr, N. 'Discussion with Einstein on epistemological problems in atomic physics'. In Paul Arthur Schlipp (ed.), *Albert Einstein: Philosopher Scientist*. Second edition. New York: Tudor Publishing Company, 1951.

Bohr, N. 'On the constitution of atoms and molecules'. *London, Edinburgh, and Dublin Philosophical Magazine*, 26 (July 1913): 1–25.

Clarke, B. R. *The Quantum Puzzle: Critique of Quantum Theory and Electrodynamics*. World Scientific Publishing, 2017.

Clifford, W. K. 'The unseen universe' [1875]. *Lectures and Essays*, Vol. 1. Cambridge University Press, 2011.

Delbruck, M. *Mind from Matter: An Essay on Evolutionary Epistemology*. Oxford: Blackwell Scientific Publications, 1986.

Dirac, P. *The Principles of Quantum Mechanics* [1930], Fourth edition. Oxford University Press, 2000.

Duhem, P. *The Aim and Structure of Physical Theory (1914)*. Translated from French by Philip P. Wiener. New York: Atheneum, 1954.

Einstein, A. *Ideas and Opinions*. New York: Bonanza Books, 1954.

Fermi, E. 'Quantum theory of radiation'. *Reviews of Modern Physics*, 4 (1932): 87–132.

Feynman, R., R. Leighton, and M. Sands (eds.). *The Feynman Lectures on Physics*, Vol. II. Pearson Addison Wesley, 2006.

Grangier, P., G. Roger, and A. Aspect. 'Experimental evidence for a photon anticorrelation effect on a beam splitter: a new light on single-photon interferences'. *Europhysics Letters*, 1 (1986): 173–179.

Gregory, R. L. 'Knowledge in perception and illusion'. *Philosophical Transactions of the Royal Society B*, 352 (1997): 1121–1128.

Häder, D.-P. 'Photosensory behaviour in procaryotes'. *Microbiological Review*, 51 (1987): 1–21.

Heisenberg, W. 'Über quantentheoretische Umdeutung kinematischer und mechanischer Beziehungen [The interpretation of kinematical and mechanical relationships according to the quantum theory]'. *Zeitschrift für Physik*, 33 (1925): 879. English translation in G. Ludwig. *Wave Mechanics*. Oxford: Pergamon Press, 1968, pp. 168–182.

Hertz, H. *The Principles of Mechanics*. English translation by D. E. Jones. London: MacMillan, 1899.

Hunt, B. *The Maxwellians*. Cornell University Press, 1991.

Larmor, J. *Aether and Matter*. Cambridge University Press, 1900.

Lodge, O. *The Ether of Space*. Harper, 1909.

Lorentz, H. A. *The Theory of Electrons*. Leipzig: B. G. Teubner, 1916.

Lorenz, K. *Behind the Mirror*. New York: First Harvest, 1978.

Lorenz, K. 'Kant's doctrine of the *a priori* in the light of contemporary biology (1941)'. English translation in H. Plotkin (ed.), *Learning Development and Culture*. Chichester: John Wiley and Sons, 1982.

Marr, D. *Vision: A Computational Investigation into the Human Representation and Processing of Visual Information*. San Francisco: W. H. Freeman, 1982.

Maxwell, J. C. 'On physical lines of force. Part I'. *The Scientific Papers of James Clerk Maxwell*. Edited by W. D. Niven, Vol. 1 (1890). New York: Dover Publications, 2003.

McMorran, B. J., A. Agrawal, I. M. Anderson, A. A. Herzing, H. J. Lezec, J. J. McLelland, and J. Unguris. 'Electron vortex beams with high quanta of orbital angular momenta'. *Science*, 331 (2011): 192–195.

Nilsson, D.-E. 'Eye evolution and its functional basis'. *Visual Neuroscience*, 30 (2013): 5–20.

Nordström, K., R. Wallén, J. Seymour, and D. Nilsson. 'A simple visual system without neurons in jellyfish larvae'. *Royal Society of London. Proceedings B. Biological Sciences*, 270 (2003), 2349–2354.

O'Connor, M., A. Garm, and D.-E. Nilsson. 'Structure and optics of the eyes of the box jellyfish'. *Journal of Comparative Physiology A*, 195 (2009): 557–569.

Piaget, J. *The Child and Reality: Problems of Genetic Psychology*. Translated by Arnold Rosin. New York: Grossman, 1973.

Piaget, J. *The Construction of Reality in the Child*. New York: Basic Books, 1971.

Popper, K. *Quantum Theory and the Schism in Physics*. First published 1982. London: Routledge, 2000.

Ryckman, T. *Einstein*. London: Routledge, 2017.

Rydberg, J. R. 'Researches sur la constitution des spectres d'émission des éléments chimiques'. *KunglSvenska Vetenskapsakademiens Handlingar*, Second series [French], 23 (1889): 1–177.

Rydberg, J. R. 'On the structure of the line-spectra of the chemical elements'. *Philosophical Magazine*, l. XXIX (1890): 331–337.

Schrödinger, E. 'What is matter?" *Scientific American*, 189(3) (1953): 52–57.

Sommerfeld, A. 'Zur quantentheorie der Spekrallinien'. *Annalen der Physik*, 51 (1916): 1–94.

Thomson, J. J. 'Cathode rays'. *The London, Edinburgh, and Dublin Philosophical Magazine*, 44(269) (1897): 293–316.

Uchida, M. and A. Tonomura. 'Generation of electron beams carrying OAM'. *Nature*, 464 (2010): 737–739.

Verbeeck, J., H. Tian, and P. Schattschneider. 'Production and application of electron vortex beams'. *Nature*, 467 (2010): 301–304.

Ziman, J. *Reliable Knowledge: An Exploration of the Grounds for Belief in Science*. First published by Cambridge University Press, 1978; Canto edition 1991.

# Part 1

# The Theory of Photons

## Chapter 2

# The Helical Space Dislocation

*Experimental evidence is presented that a photon is a single-wavelength non-rotating screw thread or helical space dislocation (HSD) that interacts with a stationary plane as circular polarization. Both the linear momentum and spin angular momentum (SAM) of light are established, not only theoretically but also in the laboratory. It is argued that SAM cannot be incorporated into a Maxwellian plane wave representation, and so the HSD model is adopted instead to derive linear momentum density, SAM density, and energy density. The correspondence between these relations and the equivalent results obtained from Maxwell's theory is presented.*

## 2.1 Circularly Polarized Rays

### 2.1.1 *Definition*

Circular polarization results from a non-rotating screw thread or helical space dislocation (HSD)[1] that is either left-wound (right-circularly polarized) or right-wound (left-circularly polarized). On passing at linear speed $c$ through a stationary plane perpendicular to its axis, it imparts a clockwise or counterclockwise rotation at speed $c$ in the plane, respectively; see Figure 2.1.[2]

The notion of a single traveling screw thread is a reasonable model for a rotating phasor on an interposed perpendicular stationary plane.

---

[1] HSD is a new acronym.

[2] Circular or linear polarization is not a property but an effect of the traveling ray. A helical spatial dislocation (HSD) — or a combination of them — projects a polarization onto a perpendicular plane that it passes through.

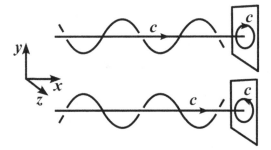

Figure 2.1   Right-circular polarization effect (top) and left-circular polarization (bottom), resulting from a HSD passing along the positive $x$ axis through a stationary plane perpendicular to the propagation direction.

### 2.1.2   *Evidence*

The idea of a transverse vibration in light rays was first suggested by Thomas Young in January 1817 after hearing a report of Fresnel's experiment in which two perpendicularly polarized rays, initially in phase, had been separated to take different paths (Young 1855, I.383). On recombination, there is no interference, whatever the chosen path difference, an effect that is assigned to their orthogonality. It was Augustin–Jean Fresnel who later that year introduced the term 'circularly polarized' to interpret his calcspar experiment; see Figure 2.2.[3]

Here, a ribbon of HSD incident along the normal of one face $A$ of a rhomb (or parallelepiped) of calcspar can be analyzed into its projected polarization components: one parallel to the incident plane (of the ribbon) — the $p$ component — and one normal to it — the $s$ component — both components being parallel to the rhomb surface, see Figure 2.2 where the two incident lines form the edges

---

[3] *"Maisquand on appliquait une feuille de papier mouillé sur une des faces réfléchissantes, le faisceau émergent paraissait complètement dépolarisé, ou polarisé circulairement, conformément au calcul.* [But when a sheet of wet paper was applied to one of the reflecting faces, the emerging beam appeared completely depolarized, or circularly polarized, according to the calculation.]" (Fresnel 1823, 430).

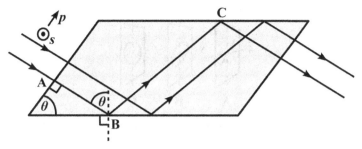

Figure 2.2 Representation of Fresnel's rhomb diagram showing that for a ribbon of rays at normal incidence on the rhomb face A, the internal angle of incidence $\theta$ at B equals the acute angle of the rhomb face (Fresnel 1823, 426). The circled dot indicates the *s* component rising out of the page, the *p* component is the arrow parallel to the incident plane.

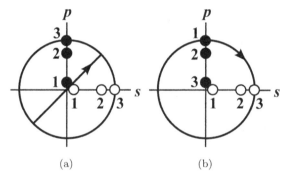

Figure 2.3 Phase relationship for a polarization effect in a stationary plane: (a) linear polarization with the p and s components in phase; (b) right-circular polarization with the *p* component advanced by $\pi/2$ radians with respect to the *s* component. The numbers along each axis indicate the order of coordinates.

of the ribbon.[4] For a refractive index of 1.51 and an internal angle of incidence $\theta = 54°37'$,[5] Fresnel found that the *p* component was advanced by $\pi/4$ radians relative to the *s* component by each internal reflection; for the result of two reflections see Figure 2.3. After two such reflections (at B and C in Figure 2.2), circular polarization

---

[4] According to Maxwell's electrodynamics, these are the components of the electric field of the ray. This is not the view to be developed here.

[5] Actually, Fresnel (1823, 425) more precisely gave $54° \ 37\frac{1}{3}'$.

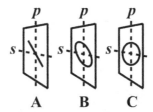

Figure 2.4   The effect of the relative phase shifts of the $p$ and $s$ components at points A, B, and C in Fresnel's rhomb (see Figure 2.2), in planes whose normals are parallel to the propagation direction of the incident rays: A has linear polarization (zero phase shift), B has elliptic polarization ($\pi/4$ relative phase shift), and C has circular polarization ($\pi/2$ relative phase shift), respectively.

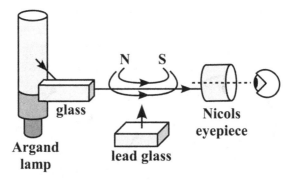

Figure 2.5   Michael Faraday's experiment on the rotation of the plane of polarization of a ray passing through a diamagnetic material permeated by magnetic field lines parallel to the ray. The lead glass is diamagnetic so the field inside the material is oppositely directed to the external field.[6]

effects emerge provided that the amplitudes of the two components are equal, an equality ensured by projecting the ribbon onto the rhomb with the linear polarization orientated at $\pi/4$ radians to the plane of incidence; see Figure 2.4.[7]

Further evidence for HSD was provided by Faraday's 1845 experiment on the magnetic rotation of linear polarization effects; see Figure 2.5. The source of his unpolarized HSD was an Argand

---

[6]Figure 2.5 is taken from Clarke (2017, 178).

[7]In September 1846, Faraday (1855, III.453–457) also reported a similar rotation effect, by passing linear polarization HSD through sugar solution without an external magnetic field.

lamp which after reflection from a glass surface at the Brewster angle produced linear polarization in a horizontal plane. By inspecting the ray through a Nichols eyepiece revolving on an axis parallel to the ray, he found that the intensity of the observed source flame depends on the angle set on the eyepiece, which can be rotated until the image is at its maximum intensity. Inserting a cuboid of silicated borate of lead between the source and the observer has no discernible effect on the observed intensity.[8] An electromagnet is now introduced into the apparatus so that the magnetic field lines are allowed to pass through the cuboid parallel (or anti-parallel) to the rays.[9] In consequence, the polarization is rotated and the eyepiece must now be adjusted to restore the image of the source flame.

> Magnetic lines, then, in passing through silicate borate of lead, and a great number of other substances, cause these bodies to act upon a polarized ray of light when the [magnetic] lines are parallel to the ray, or in proportion as they are parallel to it; if they are perpendicular to the ray, they have no action upon it. They give the diamagnetic the power of rotating the ray; and the law of this action on light is, that if a magnetic line of force be going from a north pole or coming from [to] a south pole, along the path of a polarized ray coming to the observer, it will rotate that ray to the right hand [as viewed in the ray direction]; or, if such a line of force be coming from [to] a north pole, or going from a south pole, it will rotate such a ray to the left hand. (Faraday 1855, III.5–6).

Faraday (1855, III.10–11) records having tried over 150 aqueous solutions, with varying rotation senses: for example, castor oil gave a clockwise rotation, while copaiba balm rotated the polarization counterclockwise. This variation in rotation sense for a constant field direction *when the material is varied* should emphasize the point that the magnetic lines "give the diamagnetic the power of rotating the ray" so that lines with a clockwise rotation sense (viewed toward the eyepiece) are generated by the lead glass. It follows

---

[8]Faraday's (1855, III.152) cuboid measured "2 inches square and 0.5 of an inch thick".

[9]According to Faraday (1855, 3.2153), "the poles would singly sustain a weight of from twenty-eight to fifty-six, or more, pounds".

from this opposition that the external magnetic field alone must effect a counterclockwise rotation.[10] He also found that "the rotation appears to be in proportion to the extent of the diamagnetic through which the ray and the magnetic lines pass" (Faraday 1855, III.6). No such effect was found in gases (and no effect has yet been found in vacuum), which suggests that the light rays must pass close to atomic sites where the generated reaction to the magnetic field is the strongest.

The angular displacement of the ray in radians is given by

$$\Delta\theta = VBL \tag{2.1}$$

where $B$ is the magnitude of the magnetic flux density along the ray direction, $L$ is the length of material traversed by the field and ray, and $V$ later became known as the Verdet constant for the material.[11]

### 2.1.3    *Early mathematical description*

In 1846, George Airy, the Astronomer Royal, who had been invited by Faraday to observe the rotation effect, made an attempt to cast it into mathematical form by postulating two different linear speeds for the rotations:

> I shall follow the example of Fresnel in assuming that plane polarized light may be compounded of two beams of circularly-polarized light, one right-handed and the other left-handed, and that the rotation of the plane is produced by a difference of the velocities of the two circularly-polarized beams. (Airy 1846, 470)

Here, we use right-handed axes instead of Airy's left-handed system, and change the notation for brevity. Then, Airy's displacements $Y_l$ and $Z_l$ in the $y$ and $z$ directions for a right-wound HSD, that is, a counterclockwise or left-circular polarization effect viewed in the

---

[10]The internal induced magnetic field, which is oppositely directed to the external field, produces a clockwise rotation when viewed toward the eyepiece.

[11]The Verdet constant can be wavelength dependent, and is positive for a diamagnetic material and negative for a paramagnetic one.

magnetic field direction $x$, are as follows[12]:

$$Y_l = a \cos \frac{2\pi}{\tau}\left(t - \frac{x}{v_l}\right)$$

$$Z_l = -a \sin \frac{2\pi}{\tau}\left(t - \frac{x}{v_l}\right) \qquad (2.2)$$

For a left-wound HSD or right-circular polarization effect, we have

$$Y_r = b \cos \frac{2\pi}{\tau}\left(t - \frac{x}{v_r}\right)$$

$$Z_r = b \sin \frac{2\pi}{\tau}\left(t - \frac{x}{v_r}\right) \qquad (2.3)$$

where $\tau$ is the time period of the ray, and the velocities $v_l$ and $v_r$ along the positive $x$ axis are given by

$$v_r = \left(\frac{A}{1 + \frac{C\tau}{2\pi}}\right)^{1/2}$$

$$v_l = \left(\frac{A}{1 - \frac{C\tau}{2\pi}}\right)^{1/2} \qquad (2.4)$$

Here, $A = c^2$, the square of the speed of light, so $v_r < c$ and $v_l > c$ for the constant $C > 0$. William Thomson, writing ten years later, agreed that

> Circularly polarized light transmitted through magnetized glass parallel to the line of magnetizing force, with the same quality, right-handed always, or left-handed always, is propagated at different rates [speeds] according as its course is in the direction or is contrary to the direction in which a north magnetic pole is drawn. (Thomson 1857, 152)

The difference between linear and circular polarization effects is merely a matter of phase relation between the two orthogonal

---

[12]Since Airy used a left-handed coordinate system, his signs for $Z_l$ and $Z_r$ are opposite to those given here. The rotation sense can be checked by placing the projection screen at $x = 0$ and varying the time $t$.

components, so that while for the linear form the two transverse amplitude components have a phase difference of 0 or $\pi$, for the circularly polarized form there is a $\pi/2$ or $3\pi/2$ phase difference; see Figure 2.3. The latter can be obtained from the former by using a quarter-wave plate.[13]

## 2.2   Dynamical Properties of Light

### 2.2.1   *Linear momentum*

According to Maxwell's theory, a light ray consists of two orthogonal vibrations in a plane perpendicular to the propagation axis: one being an electric, and the other a magnetic component. In 1884, John Henry Poynting suggested that

> there is a general law for the transfer of energy, according to which it moves at any point perpendicularly to the plane containing the lines of electric force and magnetic force [in the ray], and that the amount [of energy] crossing unit area per second of this plane is equal to the product of the intensities of the two forces multiplied by the sine of the angle between them divided by $4\pi$. (Poynting 1884)[14]

Starting with Maxwell's relation for the energy per unit volume $u_V$ cast in terms of $\vec{E}$ and $\vec{B}$ fields,[15]

$$u_V = \frac{1}{2}(\vec{E} \cdot \vec{D} + \vec{B} \cdot \vec{H}) \tag{2.5}$$

Poynting derives the change in this quantity per second

$$\frac{du_V}{dt} = \vec{E} \cdot \frac{d\vec{D}}{dt} + \frac{d\vec{B}}{dt} \cdot \vec{H} \tag{2.6}$$

Maxwell's displacement in a dielectric (which could be a vacuum) is proportional to the electric force $\vec{E}$ and so Poynting gives the

---

[13]For example, see Beth (1936, 115–125).

[14]Maxwell had already predicted radiation pressure 11 years earlier (1873, 2.391–392).

[15]Here, Poynting's Gaussian units are replaced by S. I. units.

displacement current as $d\vec{D}/dt$, which is the difference between the true current $\vec{S}$ and the conduction current density $\vec{J}$ so that

$$\frac{du_V}{dt} = \vec{E} \cdot (\vec{S} - \vec{J}) + \frac{d\vec{B}}{dt} \cdot \vec{H} \tag{2.7}$$

Using Maxwell's equations,

$$\frac{du_V}{dt} = -\vec{E} \cdot \vec{J} + \vec{E} \cdot (\nabla \times \vec{H}) - (\nabla \times \vec{E}) \cdot \vec{H} \tag{2.8}$$

where the $\vec{E} \cdot \vec{J}$ term is the density of electric power appearing as heat in the circuit. Making use of a vector identity and ignoring heat loss,[16] this gives the rate of change of energy in unit volume as

$$\frac{du_V}{dt} = -\nabla \cdot (\vec{E} \times \vec{H}) \tag{2.9}$$

where the amount of energy passing through unit area per second in a perpendicular plane is $\vec{E} \times \vec{H}$. In the case of electromagnetic radiation, since the $\vec{E}$ and $\vec{B}$ vectors are in the plane of propagation, the energy flow is along the propagation direction.

At the turn of the twentieth century, Nichols and Hull succeeded in showing experimentally that light carries linear momentum or radiation pressure. Their apparatus directed light from a carbon arc lamp onto silver-coated reflecting vanes suspended on a quartz fiber set in a glass container. After recording the deflection at various gas pressures, they concluded that the average radiation pressure was $1.05 \times 10^{-4}$ dynes cm$^{-2}$ ($1.05 \times 10^{-5}$Nm$^{-2}$) with a probable error of 6% (Nichols and Hull 1901, 315).[17] Short radiation exposure times were employed to reduce the action of the gas on the vanes.

---

[16]The vector identity is $\nabla \cdot (\vec{A} \times \vec{B}) = \vec{B} \cdot (\nabla \times \vec{A}) - \vec{A} \cdot (\nabla \times \vec{B})$.

[17]Maxwell (1873, 2.391–392) had already predicted that radiation pressure would occur from light rays falling on a suspended thin metallic disk in a vacuum. Adolfo Bartoli (1876, 196–202) had afterward used thermodynamics to reach the same conclusion.

## 2.2.2    *Spin angular momentum*

When John Henry Poynting (1909, 560–567) suggested that circularly polarized effects also possess angular momentum, or spin angular momentum (SAM), he used a "uniformly revolving shaft as a mechanical model of a beam of circularly polarized light". Without recourse to Maxwell's equations, he supposed his beam "to contain electrons revolving in circular orbits in fixed periodic times" (1909, 567). Given an energy per unit volume (or force per unit area) $u_V$ and wavelength $\lambda$, he found the angular momentum presented to unit area of a target surface by the beam in unit time to be $u_V \lambda / 2\pi$. Poynting also had the relation $\tan \epsilon = 2\pi a / \lambda$, where $\epsilon$ is the angle between the tangent to the helix and its axis (1909, 560), so at a rake angle of $\pi/4$, the radius of the shaft becomes

$$a = \lambda / 2\pi \tag{2.10}$$

where $\lambda$ is the pitch of the helix. In Poynting's treatment, $u_V a$ can be interpreted as the torque per unit area of a ray. In our HSD treatment, the screw thread is not rotating but moving at speed $c$ along its axis, see Figure 2.6. Since from (2.10) $\lambda = 2\pi a$ it follows that the time period is not only the time taken for the ray to progress

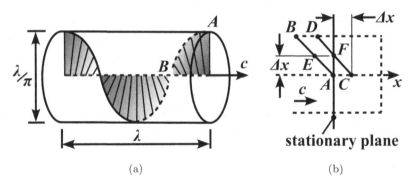

(a)                                                (b)

Figure 2.6    Representation of a HSD producing left-circular polarization passing through a perpendicular stationary plane: (a) single-wavelength right-wound screw thread or HSD with rake $\pi/4$ running along a fictitious tube at speed $c$ (B is behind A on an axis perpendicular to the page) to produce a left-circular polarization; (b) plan view of the HSD moving along the $x$ axis.

a linear distance $\lambda$ along its propagation axis, it is also the time taken for one complete rotation to be mapped out by the moving helix in a stationary plane perpendicular to the ray at a common linear and rotational speed $c$. The volume of our fictitious tube is then given by

$$V = \pi a^2 \lambda = \frac{\lambda^3}{4\pi} \qquad (2.11)$$

Although resting on a mechanical analogy, what emerges from Poynting's work and subsequent experiments is the utility of a model of light in which a right-wound or left-wound screw thread generates both linear momentum and angular momentum in a plane perpendicular to the propagation direction.

Without invoking a material medium, let us examine in more detail our representation of a circular polarization effect as the result of a HSD or non-rotating screw thread running along a fictitious guide tube at speed $c$ along the $x$ axis; see Figure 2.6. As the screw-thread blade $AB$ — with $B$ behind $A$ in Figure 2.6(a) — passes through a stationary plane perpendicular to the propagation direction — shown as a full vertical line in Figure 2.6(b) — it advances to $CD$ to create a displacement $\Delta x = AC = AF$ in time $\Delta t$, so that both the linear and azimuthal speeds have the same value $c = \Delta x / \Delta t$. Being one wavelength $\lambda$ in longitudinal and circumferential extent, the suggestion here is that this single-wavelength HSD produces an equal linear and azimuthal momentum, and an equal linear and rotational action $h$ in the stationary plane.

Using a suspended half-wave plate, Richard Beth (1936, 115–125) was the first to measure the SAM of circular polarization.[18] A quarter-wave plate is used to convert linear polarization into right- or left-circular polarization, after which the ray is passed through a half-wave plate to invert its rotation sense. In the process, a measurable angular momentum is conveyed to the plate. Certain frequency-shift experiments, most notably by P. J. Allen (1966), and Bruce Garetz and Stephen Arnold (1979), have shown that a right-circular polarization that is projected onto a plate imbued

---

[18]Details are presented in Clarke (2017, 196–201).

with a counterclockwise angular velocity $\omega_R$ is transformed into left-circular polarization with the angular velocity of the ray increasing by $2\omega_R$.[19] The angular momentum of the plate is reduced by $2\hbar$. Left-circular polarization receives a reduction of $2\omega_R$ on being transformed into right-circular polarization, while the plate angular momentum increases by $2\hbar$. Consistent with these results, Bretenaker and Floch (1990) have sent laser-generated linear polarization rays at 3.39 nm toward a rotating half-wave plate. Given $\omega$ and $\omega_R$ as the angular velocity of the wave and plate, respectively, the conclusion they reach is that "When the half-wave plate is turned with the torque of the wave, the wave produces work and loses part of its energy and vice versa". When the wave gives up energy, the plate receives angular momentum; see Table 2.1.

### 2.2.3  *The SAM plane wave problem*

The question 'Does a plane wave carry SAM?' is a crucial problem for deciding whether or not Maxwell's theory needs revising, and it is one that has been addressed by Allen and Padgett (2002, 567). The dilemma is that an infinite Maxwellian plane wave has its Poynting vector along its propagation direction and so carries no angular momentum. To do so, the Poynting vector would need an azimuthal component which implies that there would be components of the electric and magnetic vectors in the propagation direction. It would not then be a pure transverse wave. Realizing that in practice a

Table 2.1  Angular momentum and angular velocity changes for an interacting plate and ray.[20]

| Plate rotation | Polarization change | $\Delta L$ plate | $\Delta \omega$ ray |
|---|---|---|---|
| Counterclockwise | LCP to RCP | $+2\hbar$ | $-2\omega_R$ |
|  | RCP to LCP | $-2\hbar$ | $+2\omega_R$ |
| Clockwise | LCP to RCP | $-2\hbar$ | $+2\omega_R$ |
|  | RCP to LCP | $+2\hbar$ | $-2\omega_R$ |

---

[19]The latter used a half-wave quartz plate.
[20]Summary of the results due to Bretenaker and Floch (1990).

circularly polarized wave is capable of imparting SAM to a plate, the authors point out that

> even very large, uniform amplitude, diameter beams are effectively apertured by the object with which they interact. Any form of aperture introduces an intensity gradient and a detailed analysis using Maxwell's equations shows that a field component is induced in the propagation direction so that the dilemma is potentially resolved. (Allen and Padgett 2002, 567)

In other words, a target that absorbs angular momentum creates a gradient in the beam intensity $|u|^2$ at its boundary, and this is where a torque materializes. The analysis relies on an expression for the local SAM density $j_z$ as follows:

$$j_z = -\frac{r}{2}\frac{1}{|u|^2}\frac{\partial |u|^2}{\partial r}\hbar\sigma \tag{2.12}$$

where $\sigma = 0$ for linear polarized light and $\sigma = \pm 1$ for right- and left-circular polarization, respectively, with $r$ as the radial distance from the axis. The difficulty is that a Maxwellian plane wave has zero gradient and so from (2.12) it has zero spin density.

Earlier, Simmons and Gutmann (1970, 222–230) had recognized that due to the transverse nature of a Maxwellian plane wave, its linear momentum is entirely in the propagation direction and so it can have no SAM vector component in this direction.[21] This contradicts the known facts.[22] In an attempt to remedy this, they suggest that since a plane wave is unobservable, one is entitled to approximate it by introducing components of the electric and magnetic fields in the propagation direction. To facilitate the calculations, cylindrical coordinates are brought in. The authors consider the boundary of an absorbing target and suggest that only in its proximity can the electric field be non-constant. This permits non-vanishing electric and magnetic field components in the propagation direction, and

---

[21]This is due to having no azimuthal momentum as a component of the linear momentum.

[22]See also Mansuripur (2005). Here, a plane wave is *approximated* by superimposing several others that are at a non-zero angle to the propagation axis.

consequently allows an angular momentum. However, a consequence of this is that the electric field must vary with radius, but the authors find this to be inconsistent with Maxwell's $\nabla \times \vec{H} = -\partial \vec{E}/\partial t$ unless one adopts the contrary assumption that this variation can be neglected. In other words, this particular Maxwell equation breaks down at the boundary of the absorbing target where the torque from the impinging wave is to manifest itself. After obtaining the angular frequency of the radiation from the ratio of the energy density to the angular momentum density, and by the use of simplifying assumptions and an integration over the whole space, Simmons and Gutmann (1970, 230) remark that "apparent contradictions arise when we compare these theories with small volumes or at particular points in space".[23]

In fact, there are two main objections to their treatment. The first is that they assume that an impinging plane wave, which has no prior interaction with the target, possesses a region of angular momentum that somehow coincides with the boundary of the target, whatever the target's shape. This is both arbitrary and unrealistic. The second objection is that this assumption is not supported by experiment. For example, Allen (1966) has reported on a low-power microwave motor experiment in which a circular polarization ray directed at a centrally pivoted electric dipole is frequency shifted by twice the rotation frequency of the rotator. In one version of the experiment, the dipole has length 0.0165 m and the wavelength of the circular polarization ray is $\lambda = 0.032$m. So, the diameter of the radiation tube is $\lambda/\pi = 0.0102$m (see Figure 2.6) which clearly does not span the full length of the dipole. In this case, the radiation does not reach the boundary yet a torque is still conveyed from the radiation to the dipole. This contradicts the thesis of Allen and Padgett (2002).

---

[23]One simplifying assumption they make is that nowhere over the integration space does the electric field significantly vary with radius. However, if this variation can be neglected, then so can its attendant consequence, the existence of angular momentum in the direction of propagation.

An experiment conducted by Friese *et al.* (1998) has examined the transfer of torque to a trapped calcite particle. Fragments of size $1-15\,\mu\text{m}$ across and $3\,\mu\text{m}$ thick were irradiated with a $1064\,\text{nm}$ laser source which could be presented as linear or elliptical polarization rays.[24] A torque is transferred to the fragment by the latter if a photon changes its angular momentum. Again, the diameter of the radiation tube $\lambda/\pi = 3.6 \times 10^{-7}\,\text{m}$ is less than the approximate diameter of the fragment $1.5 \times 10^{-5}\,\text{m}$.[25] For the Maxwell treatment to work, a torque needs to manifest at the boundary, but there are observed cases where it does not. Clearly, there needs to be a new model of a wave carrying SAM.

## 2.3 The HSD Model

### 2.3.1 *Spin-1*

As we have already seen, a 'single photon' involves a HSD or non-rotating screw thread, one wavelength in length both along its propagation axis and around the circumference of its cross-section,[26] which projects motion onto a stationary plane at speed $c$ both linearly and azimuthally.

Let us define 'spin-1' as the rotation induced in a stationary plane set perpendicular to the propagation axis by a HSD. In what follows, for all variables relating to spin-1, a '1' appears in the subscript, for example, $r_1$ for the tube radius. For circular polarization HSD in the $y-z$ plane, let the radius $\vec{r}_1$ and the azimuthal velocity $\vec{v}_1$ together with their complex conjugates be given by (Clarke 2017, 208–209)

$$\vec{r}_1 = r_1 \exp(\mathrm{i}\omega t) \begin{pmatrix} \pm\mathrm{i} \\ 1 \end{pmatrix}$$

$$\vec{r}_1^* = r_1 \exp(-\mathrm{i}\omega t) \begin{pmatrix} \mp\mathrm{i} \\ 1 \end{pmatrix} \tag{2.13}$$

---

[24]In the first experiment, the fragment acted as a half-wave plate and rotated the plane of linear polarization. The second experiment used elliptical polarization rays obtained from sending linear polarization rays through a quarter-wave plate.
[25]These values are given in error by Clarke (1997, 207).
[26]Measured in a plane perpendicular to its axis.

and

$$\vec{v}_1 = \frac{d\vec{r}}{dt} = i\omega r_1 \exp(i\omega t) \begin{pmatrix} \pm i \\ 1 \end{pmatrix}$$

$$\vec{v}_1^* = \frac{d\vec{r}^*}{dt} = -i\omega r_1 \exp(-i\omega t) \begin{pmatrix} \mp i \\ 1 \end{pmatrix} \tag{2.14}$$

where throughout the analysis that follows, the top sign in the $y$ component (upper component of the $y - z$ column vector) is for right-circular and the bottom sign is for left-circular polarization as viewed in planes in the direction of propagation; see Figure 2.7.[27] With $r_1$ as the maximum radius (which Poynting expressed as $a$), the following relations arise:

$$\vec{r}_1^* \times \vec{r}_1 = r_1^2 \exp(-i\omega t) \exp(i\omega t) \begin{pmatrix} \mp i \\ 1 \end{pmatrix} \times \begin{pmatrix} \pm i \\ 1 \end{pmatrix} = \mp 2i \vec{r}_1^2 \hat{\imath}$$

$$\vec{r}_1^* \times \vec{v}_1 = i\omega r_1^2 \exp(-i\omega t) \exp(i\omega t) \begin{pmatrix} \mp i \\ 1 \end{pmatrix} \times \begin{pmatrix} \pm i \\ 1 \end{pmatrix} = \pm 2\omega \vec{r}_1^2 \hat{\imath}$$

$$\vec{v}_1^* \times \vec{v}_1 = \omega^2 r_1^2 \exp(-i\omega t) \exp(i\omega t) \begin{pmatrix} \mp i \\ 1 \end{pmatrix} \times \begin{pmatrix} \pm i \\ 1 \end{pmatrix} = \mp 2i\omega^2 r_1^2 \hat{\imath}$$

$$\tag{2.15}$$

where $\hat{\imath}$ is a unit vector in the propagation direction $x$.

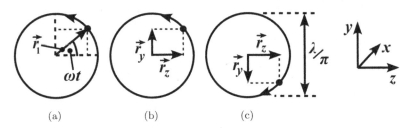

(a)            (b)            (c)

Figure 2.7    Complex number representation of Equation (2.13) on a stationary plane, for a circular polarization HSD viewed in the propagation direction $x$: (a) dependence of phasor $\vec{r}_1$ on $\omega t$; (b) $\vec{r}_1$ components of left-circular polarization with $y$ component as $-i$; (c) $\vec{r}_1$ components of right-circular polarization with $y$ component as $+i$.

---

[27]Note that the $i$ in the $y - z$ column vector can be written as $\exp(i\pi/2)$.

### 2.3.2 *Dynamical results*

In what follows, where $\pm$ or $\mp$ occurs, the top sign is for right-circular polarization and the bottom sign is for the left.

- Spin angular momentum

  Let the SAM be defined by

$$\vec{J} = \frac{k_1}{2}\vec{r}_1^* \times \vec{v}_1 = \pm k_1 \omega r_1^2 \hat{\imath} \tag{2.16}$$

  where $k_1$ is constant. The magnitude of the SAM for a photon must be $\hbar = k_1 \omega r_1^2$, thereby establishing the constant $k_1$.

- Linear momentum

  Let us define the linear momentum with

$$\vec{p} = \frac{ik_1}{2c}\vec{v}_1^* \times \vec{v}_1 = \frac{ik_1}{2c}\left(\mp 2i\omega^2 r_1^2\right)\hat{\imath} = \pm k_1 \omega r_1 \hat{\imath} = \pm \frac{\hbar}{r_1}\hat{\imath} \tag{2.17}$$

  having used $k_1$ from (2.16) and noting that the rotation in the stationary plane occurs at speed $\omega r_1 = c$. Experimentally, the linear momentum direction is independent of the spin-1 sense so a magnitude sign could be employed here.

- Energy of a single tube

  Let the energy $\varepsilon$ carried by a single-wavelength HSD tube be defined by

$$\varepsilon = \vec{p} \cdot \vec{c} \tag{2.18}$$

  where $\vec{c} = \pm c\hat{c}$ is the velocity of light in a given direction $\hat{c}$. If this is the propagation direction $\hat{\imath} = \hat{c}$, then from (2.17), since $\omega r_1 = c$, we have

$$\varepsilon = \hbar\omega \tag{2.19}$$

  The following relations now introduce the volume $V$ of a single tube.

- Spin angular momentum density

  The volume of a HSD tube is given from (2.11), so using (2.15) and (2.16) we have the SAM volume density $\vec{J}_V$ of a tube as

  $$\vec{J}_V = \frac{k_1}{2V} \vec{r}_1^* \times \vec{v}_1 = \pm \frac{4\pi\hbar}{\lambda^3} \hat{\imath} \qquad (2.20)$$

- Linear momentum density

  Dividing (2.17) by $V$ and using (2.11) leads to the linear momentum density

  $$\vec{p}_V = \frac{ik_1}{2cV} \vec{v}_1^* \times \vec{v}_1 = \pm \frac{4\pi h}{\lambda^4} \hat{\imath} \qquad (2.21)$$

  where $h$ is Planck's constant, having used (2.10) with $\lambda = 2\pi r_1$.

- Energy density

  Using (2.19) with (2.11), the energy density is

  $$u_V = \frac{\varepsilon}{V} = \frac{4\pi\hbar\omega}{\lambda^3} \qquad (2.22)$$

We now investigate the correspondence between HSD equations (2.16)–(2.22) and those relations that arise from the assumption that the transverse components of a light ray are electric field vectors. When considering electric and magnetic fields, it should be kept in mind that a magnitude squared is a volume density, so that our interest is confined to (2.20)–(2.22). Also, the magnetic field vector in a plane wave front is perpendicular to the electric field vector and is related in magnitude by $\left|\vec{B}\right| = \left|\vec{E}\right|/c$. Nevertheless, it need not play a part at all.[28]

---

[28] "The polarization of a plane light wave propagating along the $z$ direction is defined by a pair of electric field components $E_x$ and $E_y$ arranged as a column vector $\vec{E}$"; see Simon *et al.* (1988, 19).

### 2.3.3   *Correspondence to Maxwell's theory*

Let us define the electric field phasor as $\vec{E} = -k_2\vec{r}_1$, where $k_2$ is a positive constant. Then, using (2.13) and (2.14), we have

$$\vec{r}_1 = -\frac{\vec{E}}{k_2} = -\frac{E_o}{k_2}\exp(i\omega t)\begin{pmatrix}\pm i\\1\end{pmatrix}$$

$$\vec{r}_1^* = -\frac{\vec{E}^*}{k_2} = -\frac{E_o}{k_2}\exp(-i\omega t)\begin{pmatrix}\mp i\\1\end{pmatrix} \qquad (2.23)$$

and

$$\vec{v}_1 = \frac{d\vec{r}_1}{dt} = -\frac{1}{k_2}\frac{d\vec{E}}{dt} = -\frac{i\omega E_o}{k_2}\exp(i\omega t)\begin{pmatrix}\pm i\\1\end{pmatrix}$$

$$\vec{v}_1^* = \frac{d\vec{r}_1^*}{dt} = -\frac{1}{k_2}\frac{d\vec{E}^*}{dt} = \frac{i\omega E_o}{k_2}\exp(-i\omega t)\begin{pmatrix}\mp i\\1\end{pmatrix} \qquad (2.24)$$

having placed $r_1 = -E_o/k_2$.

- Spin angular momentum density
  We now adopt a prime on a variable to denote a Maxwellian volume density. Using (2.20), (2.23), and (2.24), we have

$$\vec{J}_V' = \frac{k_1}{2V}\vec{r}_1^* \times \vec{v}_1 = \frac{k_1}{2Vk_2^2}\vec{E}^* \times \frac{d\vec{E}}{dt} = \pm\frac{k_1\omega E_o^2}{k_2^2 V}\hat{\imath} \qquad (2.25)$$

Now, the SAM density is usually given as follows (Friese *et al.* 1998, 348):

$$\vec{J}_V' = \frac{\epsilon}{2i\omega}\vec{E}^* \times \vec{E} \qquad (2.26)$$

and using $\vec{E}_o = -k_2\vec{r}_1$, (2.13) and (2.14) in (2.26) we have

$$\vec{J}_V' = \frac{\epsilon k_2^2}{2i\omega}\vec{r}_1^* \times \vec{r}_1 = -\frac{\epsilon k_2^2}{2\omega^2}\vec{r}_1^* \times \vec{v}_1 \qquad (2.27)$$

Since $k_1 = \hbar/\omega r_1^2$, then comparing the coefficients of $\vec{r}_1^* \times \vec{v}_1$ in (2.25) and (2.27), taking $V$ from (2.11), and using $\lambda = 2\pi c/\omega =$

$2\pi r_1$, we arrive at

$$k_2^2 = -\frac{\hbar\omega}{\epsilon r_1^2 V} = -\frac{2\pi^2 \hbar c}{\epsilon V^2} \tag{2.28}$$

- Linear momentum density

From (2.21) and (2.24), we find

$$\vec{p}_V' = \frac{ik_1}{2cV}\vec{v}_1^* \times \vec{v}_1 = \pm\frac{k_1\omega^2 E_o^2}{ck_2^2 V}\hat{\imath} \tag{2.29}$$

- Energy density

If the direction $\hat{c}$ in which the energy is collected is that same as the momentum propagation direction $\hat{\imath}$, then (2.18) and (2.29) gives

$$u_V' = \vec{p}_V' \cdot c\hat{\imath} = \pm\frac{k_1\omega^2 E_o^2}{k_2^2 V} = \mp\epsilon E_o^2 \tag{2.30}$$

We already have $k_1 = \hbar/\omega r_1^2$. The power per unit area is given by the Poynting vector

$$\vec{S} = \epsilon c^2 \vec{E} \times \vec{B} = \frac{\epsilon c}{2i}\vec{E}^* \times \vec{E} = \mp\epsilon c E_o^2 \hat{\imath} \tag{2.31}$$

From (2.31), if we consider the relative directions of the Poynting vector $\hat{S}$ and the target surface normal $\hat{\imath}$, then the absorbed energy density is

$$u_V' = \frac{\vec{S} \cdot \hat{\imath}}{c} = \mp\epsilon E_o^2 \tag{2.32}$$

Using (2.23), (2.24), and $\vec{r}_1 = \vec{v}_1/i\omega$ finally leads to

$$\vec{E} = -k_2\vec{r}_1 = \pm i\sqrt{\frac{2\pi^2 \hbar c}{\epsilon V^2}}\vec{r}_1 = \pm\frac{1}{\lambda^2}\sqrt{\frac{4\pi h}{\epsilon c}}\vec{v}_1 \tag{2.33}$$

where $h$ is Planck's constant. Here, the $\vec{E}$ vector in the light ray becomes an expression of the circular polarization velocity phasor $\vec{v}_1$. Despite the corroboration initiated by Hertz of the Maxwellian theory of light propagation, the difficulties it has in representing SAM presents itself as an invitation to relinquish electric field vectors as components of light rays and replace them with HSD, which induce displacements in a stationary plane. There is also a deeper philosophical reason for this abandonment. After all, it would be better to rest the theory of light structure, and of fields in general, on a geometrical phasor representation using $\vec{r}_1$ or $\vec{v}_1$, which occurs naturally in consideration of circular polarization effects. At present it is grounded on an electric field vector that has no geometrical basis. We should also keep in mind that Maxwell's theory falls short in accounting for both atomic stability and the theory of blackbody radiation and so it cannot possibly be a reliable theoretical basis.

This shows how Maxwellian relations such as those for SAM density, linear momentum density, and energy density can arise out of a theory describing a directed HSD occupying a volume $V$. The next task is to examine the evidence that a single-photon wave front is capable of more than one excitation, for this strikes at the very heart of the quantum mechanics paradox.

## References

Airy, G. B. 'On the equations applying to light under the action of magnetism'. *The London, Edinburgh and Dublin Philosophical Magazine*, 28 (1846): 469–477.

Allen, L., and M. J. Padgett. 'Does a plane wave carry spin angular momentum?' *American Journal of Physics*, 70 (2002): 567–568.

Allen, P. J. 'A radiation torque experiment'. *American Journal of Physics*, 34 (1966): 1185–92.

Bartoli, A. 'Il calorico raggiante e il secondo principio di termodynamica'. *Nuovo Cimento*, 15 (1876): 196–202.

Beth, R. A. 'Mechanical detection and measurement of the angular momentum and light'. *Physical Review*, 50 (1936): 115–125.

Bretenaker, F., and A. Le Floch. 'Energy exchanges between a rotating retardation plate and a laser beam'. *Physical Review Letters*, 65 (1990): 2316–2317.

Clarke, B. R. *The Quantum Puzzle: Critique of Quantum Theory and Electrodynamics*. Singapore: World Scientific Publishing, 2017.

Faraday, M. 'On the magnetic affection of light', *Experimental Researches*, 3 vols. London: Taylor and Francis, 1855.

Fresnel, M. A. 'Mémoire sur la loi des modifications que la réflexions imprime a la lumière polarisée [Memoir on colors developed in homogeneous fluids by polarized light] (1817)'. *Mémoires de l'Académie Royale des Sciences*, 11 (Paris, 1823), pp. 393–433.

Friese, M. E. J., T. A. Nieminen, N. R. Heckenberg, and H. Rubinsztein–Dunlop. 'Optical alignment and spinning of laser-trapped microscopic particles'. *Nature*, 394 (1998): 348–350.

Garetz, B. A., and S. Arnold. 'Variable frequency shifting of circularly polarized laser radiation via a rotating half-wave retardation plate'. *Optical Communications*, 31 (1979): 1–3.

Mansuripur, M. 'Angular momentum of circularly polarized light in dielectric media'. *Optics Express*, 13 (2005): 5315–5324.

Maxwell, J. C. *A Treatise on Electricity and Magnetism*, 2 vols. Oxford: Clarendon Press, 1873.

Nichols, E. F. and G. F. Hull. 'A preliminary communication on the pressure of light and heat'. *Physical Review*, 13 (1901): 307–320.

Poynting, J. H. 'On the transfer of energy in the electromagnetic field'. *Philosophical Transactions of the Royal Society of London*, 175 (1884): 343–361.

Poynting, J. H. 'The wave motion of a revolving shaft, and a suggestion as to the angular momentum in a beam of circularly polarized light'. *Proceedings of the Royal Society of London A*, 82, 557 (1909): 560–567.

Simmons, J. W., and M. J. Gutmann. *States, Waves, and Photons: A Modern Introduction to Light*. Reading, MA: Addison–Wesley, 1970.

Simon, R., H. J. Kimble, and E. C. G. Sudarshan. 'Evolving geometric phase and its dynamical manifestation as a frequency shift: An optical experiment'. *Physical Review Letters*, 61 (1988): 19–22.

Thomson, W. 'Dynamical illustrations of the magnetic and helicoidal rotary effects of transparent bodies on polarized light'. *Proceedings of the Royal Society of London*, Vol. 8. London: Taylor and Francis, 1857, pp. 150–157.

Young, T. *Miscellaneous Works of the late Thomas Young*, 2 vols. London: John Murray, 1855.

Chapter 3

# A Transversely Iterated Single Photon

*The classic Grangier et al. (1996) experiment is revisited to suggest an alternative model of a single photon. Experiments using spontaneous parametric down conversion (SPDC) show that not every single-photon emission triggers a detector and that multiple triggering can occur during a single gate. The latter is usually interpreted as being caused by a spurious photon entering the gate from a different SPDC event. However, an alternative model is suggested in which there are no spurious events, and a single photon manifests as a transversely iterated front of helical space dislocations (HSDs).On exiting a beam splitter, it can produce multiple triggering of detectors with a probability of $1.0 \times 10^{-2} - 7.5 \times 10^{-4}$. Evidence in the work of Steinberg et al. (1992) is also reinterpreted to suggest that a photon is not a wave packet.*

## 3.1 Attenuated Beams

Even before reliable experimental evidence became available, Dirac (2000, 9) had pointed out that "Interference between two photons never occurs". A photon always interferes with itself. Early experiments were designed to isolate a single photon by employing various obstacles to reduce the intensity of an aggregate beam of photons. Following Dirac's suggestion, the British physicist Geoffrey Taylor set out to show that a photon indeed interferes with itself rather than with another photon. After reducing the intensity of light from a gas source by passing it through a narrow slit followed by smoked glass screens, Taylor (1909, 114–115) diffracted it around a sewing needle onto a photographic plate to demonstrate that interference could occur. He estimated the power of the beam falling on $1\,\text{cm}^2$ to be

$5 \times 10^{-6}$ ergs s$^{-1}$, but since a single photon of red light has an energy of $2.7 \times 10^{-12}$ ergs, a calculation shows that approximately $10^6$ photons were striking 1 cm$^2$ of his photographic plate per second.

The production of single photons was refined by Dempster and Batho (1927, 644) who used a blackbody as reference to estimate that the energy radiated in the helium line 4,471 Å from their low-pressure glow discharge amounted to an average of 95 photons s$^{-1}$. From the product of the transition time ($5 \times 10^{-8}$ s) and the speed of light in vacuum, they calculated the minimum spatial separation between successive emissions from a single source to be 15 m yielding an average density of about 6 photons m$^{-1}$. They concluded that to produce the interference effects they observed, a single photon must cover a wave-front area of at least 32 mm$^2$. In a second experiment, they separated an individual photon into components using a half-silvered mirror and then recombined it with a relative phase difference.

Reynolds *et al.* (1969, 355) managed to improve on the experiment by investigating whether or not the quality of the interference pattern diminishes as the light intensity approaches that of a single photon. In two experiments, one with a mercury discharge tube and the second with an RF excited low-pressure Hg lamp, they estimated that their Fabry–Perot interference patterns were formed from a beam of 15 photons s$^{-1}$ and 30 photons s$^{-1}$, respectively. Confident that no more than one photon was ever engaged in an interference interaction, they reported no evidence that the contrast of their interference pattern was diminished by the use of single photons.[1] In more recent times, an electron microscope fitted with an electron biprism has been used to demonstrate the accumulation of an interference pattern from a single electron. The experimenters suggest that "there is very little chance for two electrons to be present simultaneously between the source and the detector, and much less chance for two wave packets to overlap" (Tonomura *et al.* 1989, 119).

---

[1] An experiment by Dontsov and Baz (1967, 1) found a deterioration in contrast, but attempts to reproduce their findings have met with failure. For a review of single-photon experiments, see Pipkin (1979, 294).

## 3.2    Coincidence Counting

Grangier *et al.* (1986, 174) have expressed dissatisfaction with methods that rely on strongly attenuated beams, suggesting that "none has been performed with single-photon states of light. As a matter of fact, all have been carried out with chaotic light". Earlier, Philippe Grangier had collaborated with Alain Aspect and Gérard Rogier to measure "the linear polarization correlation of the photons emitted in a radiative atomic cascade of calcium" selectively pumped to the upper level of the cascade (Aspect *et al.* 1981, 460). In the process, two photons are produced from the transitions $4p^2\,{}^1S_0 - 4s4p\,{}^1P_1 - 4s^2\,{}^1S_0$, correlated in polariza-tion, with wavelengths $\lambda_1 = 551.3\,\text{nm}$ and $\lambda_2 = 442.7\,\text{nm}$; see Figure 3.1. The photons are sent in opposite directions through aspheric lenses ($f = 40\,\text{mm}$, diameter $= 50\,\text{mm}$), followed by a set of 10 optically flat polarization plates inclined near to Brewster's angle, and then through filters to remove unwanted wavelengths. They eventually arrive at the photomultiplier detectors P.M. 1 and P.M. 2 where the excitation-event signaling is routed into coincidence counting electronics. The time delay between coincident detections is measurable and results are recorded around the null delay in a 19 ns coincidence window. The authors have reported singles counting rates of "40,000 and 120,000 counts per second" for $\lambda_1$ and $\lambda_2$, respectively, and "coincidence rates without polarizers [of]

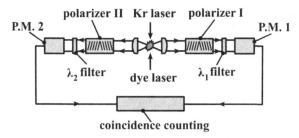

Figure 3.1    Schematic diagram of apparatus for counting coincidence detections between two simultaneously generated photons (Aspect *et al.* 1981). The laser beams are focused onto the atomic beam which runs perpendicular to the figure. Filters ensure that the photomultipliers (P.M.) only receive photons of wavelength $\lambda_1$ at P.M. 1 and wavelength $\lambda_2$ at P.M. 2.

240 coincidences per second" (Aspect *et al.* 1981, 461–62).[2] The probability $P_C$ of detecting a coincidence given that a detector has been excited can be approximated from

$$P_C = \frac{B}{A + B} \tag{3.1}$$

where $A$ is the total singles count taken over both detectors and $B$ is the coincidence count. From this, we find that $P_C = 1.5 \times 10^{-3}$. The fact that a photon pair is generated means that $P_C$ can also be interpreted as the probability that a particular detector registers an absorption given that a photon is incident on it, known from the registration of the photon at the other detector.

Similar experiments have also been performed using spontaneous parametric down conversion (SPDC) to generate equal-frequency photon pairs — the signal and idler — which are each sent to their own detector to obtain singles and coincidence counts (Noh *et al.* 2006; Ahmed *et al.* 2014); see Table 3.1.[3] Noh *et al.* (2006) remark that the singles count rates can vary with the focal length of the aspheric lens in the fiber optic collimators and the 'accidental' coincidence counts can vary with the laser output power, and although 'accidental' counts are understood to be fewer than 'true' ones, the authors omit details as to how they are to be distinguished. Kwiat *et al.* (1999) have stated that "typical 'accidental' coincidence rates are negligible $(<1\,\mathrm{s}^{-1})$" and attempt no explanation for them. Their experimental arrangement used half-wave plates in both the idler and signal branches in order to rotate the plane of polarization.

---

[2]The authors estimate "accidental coincidences" by counting coincidences $100\,\mathrm{ns}$ after the null delay and have found 90/s. So, a subtraction yields 150 "true coincidences per second".

[3]SPDC is a quantum optical process where a nonlinear $\chi^{(2)}$ crystal is used to down convert a pump photon of higher energy into a pair of photons of lower energy. A continuous or pulsed laser source is used to supply the nonlinear crystal in which energy and momentum are conserved. Type I phase matching is where the photons emerge with the same linear polarization, while Type II results in mutually perpendicular (polarization-entangled) polarizations. The latter process tends to produce fewer photons than the first.

Table 3.1 Estimate of the probability of a detector registration given that a photon is incident on it, which is known from the detection of its correlated conjugate.

| $P_C$ | A | B | Source |
|---|---|---|---|
| $1.5 \times 10^{-3}$ | $160{,}000\,\mathrm{s}^{-1}$ | $240\,\mathrm{s}^{-1}$ | Aspect *et al.* (1981) |
| $1.17 \times 10^{-3}$ | $8{,}500\,\mathrm{s}^{-1}$ | $9.96\,\mathrm{s}^{-1}$ | Grangier *et al.* (1986)[4] |
| $1.38 \times 10^{-2}$ | $1{,}500{,}000\,\mathrm{s}^{-1}$ | $21{,}000\,\mathrm{s}^{-1}$ | Kwiat *et al.* (1999)[5] |
| $1.03 \times 10^{-2}$ | $2{,}800\,\mathrm{s}^{-1}$ | $29\,\mathrm{s}^{-1}$ | Noh *et al.* (2006)[6] |
| $7.494 \times 10^{-4}$ | $200{,}000\,\mathrm{s}^{-1}$ | $150\,\mathrm{s}^{-1}$ | Ahmed *et al.* (2014)[7] |

Using Equation (3.1), counts for five experimental studies are summarized in Table 3.1 from which it should be clear that when a photon is incident on a detector, known from the detection of its conjugate, then its presence is not always registered.

## 3.3 Beam Splitting the Signal Photon

If a photon front can impinge on a detector without necessarily affecting a registration, we now ask the following: Is it possible for a single-photon front to be divided at a beam splitter into two components, each of which can produce a registration at their respective detectors as a coincidence?

The crucial experiment was performed by Grangier *et al.* (1986) who set up a double-photon source (atomic cascade) in which the

---

[4]The signal beam was divided at the beam splitter. On subtracting dark rates, each beam splitter detector registered half the rate given under $B$ in Table 3.1. The coincidence count is calculated using the sum of the two beam-splitter divisions which is $2 \times 4.98\,\mathrm{s}^{-1}$.

[5]Two relatively thin BBO nonlinear crystals cut for Type I were set adjacent to each other to create a pair of degenerate photons at 702 nm. Their bandwidth was 5 nm.

[6]A lithium triborate (LBO) nonlinear crystal cut for Type I was used to yield two photons at 1,550 nm each.

[7]A $BiB_3O_6$ (BiBO) nonlinear crystal cut for Type I was used to create two photons at 804 nm each. Dr. Mubarack Ahmed has assured me by email that the value $200{,}000\,\mathrm{s}^{-1}$ is the total singles count.

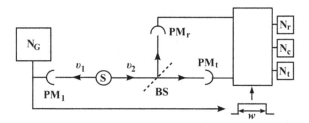

Figure 3.2 Triggered anticorrelation experiment. The detection of the first photon of the cascade at $PM_1$ produces a gate $w$, during which time the photomultipliers $PM_t$ and $PM_r$ are active. The probabilities of detection during the gate are $p_t = N_t/N_G$, $p_r = N_r/N_G$ for singles, and $p_c = N_c/N_G$, for coincidences. Grangier *et al.* (1986; Figure 1, 174).

photons emerge with different frequencies. One precedes the other by a known time interval, and the first is used to trigger the detection apparatus for the second. Two experiments are then carried out, both with the same source and rate counters. The first runs the source S photons into a beam splitter BS, without recombination of the fronts, with a photomultiplier PM and counter on each output path to bring out the quantum mechanical or individual nature of the photons in an anti-correlation; see Figure 3.2.[8] The second replaces the beam splitter with a Mach–Zender interferometer, with adjustable path differences, a device that is intended to bring out the classical or relative phase relationship between the two paths in a correlation; see Figure 3.3. In contrast to the second case, the first is expected to raise no coincidences between the paths $\nu_2$ can take. Here, the authors assert that "a single-photon can be detected once [only]" and cite their result that they measure coincidences that are "five times smaller than the classical [wave theory] lower limit" with a "maximum violation of more than 13 standard deviations" for the classical coincidence inequality.[9] However, their statement is in need of qualification. The fact is, they actually measure some coincidences. Their low count does not mean that coincidences *do not* occur.

---

[8]This was first performed by Hanbury Brown and Twiss (1956a).
[9]The expected number of coincidences from the classical theory should be $\geq 50$, whereas they measure only 9.

It means that they *rarely* occur, and it is suggested that they cannot be dismissed as mere errors in the experiment.

For the first Grangier experiment (Figure 3.2), two photons with frequencies $v_1$ and $v_2$ are generated by a cascading process at the source S, with the decay of the intermediate state of the cascade having a lifetime of $\tau_s = 4.7$ ns. The detection of $v_1$ at the photomultiplier $PM_1$ triggers a gate that allows the photomultipliers $PM_r$ and $PM_t$ to be exposed to the photon with frequency $v_2$ for a time interval $w \approx 2\tau_s$.[10] Singles and coincidence counters are fed by the photomultipliers, and given that $\Delta T$ is the counting interval, then $N_G/\Delta T$ is the rate at which the gate is triggered, $N_r/\Delta T$ is the reflected singles rate for $PM_r$, $N_t/\Delta T$ is the transmitted singles rate for $PM_t$, and $N_c/\Delta T$ is the coincidence rate for a registration by both $PM_r$ and $PM_t$ during a single gate. It was found that during $1.53 \times 10^8$ gate events, $N_t = N_r = 89{,}640$ singles were counted at $PM_t$ and $PM_r$ with only $N_c = 9$ coincidences recorded.[11] Applying calculation (3.1), we find $P_C = 5.02 \times 10^{-5}$. In consequence, their report declares, "Hence the light emitted after each triggering pulse has been shown to exhibit a specifically quantum anticorrelation behaviour" (Grangier *et al.* 1986, 177).[12] Since this part of the experiment involves only a splitting of the beam at BS with no reconstitution, no opportunity is afforded for components of the beam to engage in interference. Although a photon is presently understood to take one path only out of the beam splitter, the occurrence of 9 coincidences allows the interpretation that this is not always the case. These coincidences have actually been observed.

The Grangier experiment was repeated 18 years later using SPDC with a 2.5 ns gate over a 5-minute time period (Thorn *et al.* 2004,

---

[10]The excitation of $PM_1$ triggers the start of the gate, and the excitation of either $PM_r$ or $PM_t$ signals its termination.

[11]The total counting time was $T = 5$ h, and the $PM_1$ rate $N_G$ was $8{,}800$ s$^{-1}$ which includes a dark rate (no cascade source) of $300$ s$^{-1}$. The rate $N_r$ was $5$ s$^{-1}$ which includes a dark rate of $0.02$ s$^{-1}$. In Table 3.2, dark rates are subtracted. 'Dark rate' means a count with no source except the laboratory environment.

[12]'Anticorrelation' means that the sum of the two photon energies is constant.

Table 3.2    The probability of a second post-BS detector registering given that one of them has already registered during a gate.

| Coincidence probability given photon incidence | Source |
| --- | --- |
| $5.02 \times 10^{-5}$ | Grangier *et al.* (1986) |
| $3.54 \times 10^{-4}$ | Thorn *et al.* (2004) |

1210–1219).[13] A 'gate opening' occurs by the excitation of the idler detector, and the rate of idler detection was $R_G \sim 110{,}000\,\text{s}^{-1}$. So, the three-fold coincidence rate $R_c$ obtained by Thorn *et al.* (2004), which is not stated, can be found from their best calculated value of $\alpha = R_c R_G/(R_r R_t) = 0.0177 \pm 0.0026$, where $R_r = R_t = 4{,}400\,\text{s}^{-1}$ at $\text{PM}_r$ and $\text{PM}_t$, from which it appears that they actually measured $R_c = 3.12$ counts/s.[14] Given that a single signal emission has registered at a detector $\text{PM}_r$ or $\text{PM}_t$ or both, the probability of coincidence is then $P_c = 3.12/(2 \times 4400 + 3.12) = 3.54 \times 10^{-4}$. In Table 3.2, we see the coincidence rates from the two studies mentioned. Thorn *et al.* (2004) argue that the 'real' coincidence rate for spatially localized photons should be $P_c = 0$ and speculate that these 'accidental coincidences' are "uncorrelated photons from different down-conversion events [that] may hit the T and R detectors within our finite coincidence window" (Thorn *et al.* 2004, Appendix A). They suggest that whenever a coincidence occurs during a gate, an intruder-photon is passing through the beam splitter from a different SPDC event. It should be emphasized that they have no way of observing this, that their 'explanation' is actually a conjecture, and in that case there is space for an alternative one. The alternative suggested here is that a single-photon front incident on more than one

---

[13]Since the dark count rate was $250\,\text{s}^{-1}$, this produces negligible $6.25 \times 10^{-7}$ dark effects per gate.

[14]This was obtained from a 40-min count. A three-fold coincidence is an idler detection that starts the gate with a signal detection registered at both $R$ and $T$ during the gate. Thorn *et al.* (2004) use $R_G$, whereas Grangier *et al.* (1986) use $R_1$.

detector can produce more than one registration. From Table 3.2, its probability of doing so appears to be about $5.0 \times 10^{-5} - 3.5 \times 10^{-4}$.[15]

## 3.4   Recombination of the Beam Splitter Outputs

So far, we have looked at the probability that given a single-photon front is incident on a detector, it effects a registration (Section 3.1), and the probability that given a single-photon front affects one detector, it also registers at another (Section 3.2). We now turn to the second part of the Grangier experiment which makes use of the same triggering mechanism. The opportunity for a superposition of parts of the signal-photon front is now available. Here, a Mach–Zender interferometer replaces the beam splitter BS and the photomultipliers $PM_r$ and $PM_t$; see Figure 3.3. The interferometer operates as follows.

### 3.4.1   *Principles of reflection and transmission*

(1) A half-silvered mirror reflects half of the incident light and refracts the other half.
(2) A $\pi$ phase shift results when a ray is reflected from a surface with a denser medium on the other side (e.g., air to glass).

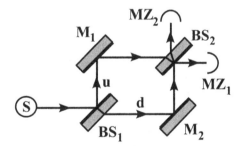

Figure 3.3   A Mach–Zender interferometer. The detection probabilities in outputs $MZ_1$ and $MZ_2$ are oppositely modulated as a function of the path difference between the arms of the interferometer, varied at $u$. The thick lines on the beam splitters $BS_1$ and $BS_2$ indicate reflection surfaces, and $M_1$ and $M_2$ are mirrors; see Grangier *et al.* (1986; Figure 3, 177).

---

[15] From Table 3.1, the range for a single registration at a detector is $7.5 \times 10^{-4} - 1.0 \times 10^{-2}$. Of course, the detector construction affects its absorption efficiency.

(3) No phase shift results when a ray is reflected from a surface with a less dense medium on the other side (e.g., glass to air).

(4) A mirror causes a $\pi$ phase shift.

(5) When a ray passes through a medium, there is a phase change, the magnitude of which depends on the refractive index of the medium and the path length.

There are two possible paths for a photon to take on exiting the first beam splitter $BS_1$: u (up), or d (down), where path u has variable length. Table 3.3 shows the relative phase shifts between these two paths for detectors $MZ_1$ and $MZ_2$.

We can see from Table 3.3 that when $\delta = 0$, there is constructive interference at $MZ_1$ but a destructive one at $MZ_2$, and vice versa when $\delta = \pi$. Grangier *et al.* (1986) report varying the path difference $\delta$ in 256 steps of increment $\lambda/50$, with a counting time of 1 s per stop. The accumulated count for 15 sweeps is shown in Figure 3.4. The

Table 3.3    Relative phase shift between the u and d paths in Figure 3.3 for a Mach–Zender interferometer measured at detectors $MZ_1$ and $MZ_2$ for a path difference $\delta$ (rads).

| Destination | Path u | Path d | Relative phase shift |
|---|---|---|---|
| Detector $MZ_1$ | $BS2(\pi)$, $\delta$, $M4(\pi)$, BS5 | BS5, $M4(\pi)$, $BS2(\pi)$ | $\delta$ |
| Detector $MZ_2$ | $BS2(\pi)$, $\delta$, $M4(\pi)$, BS5, BS3(0), BS5 | BS5, $M4(\pi)$, BS5 | $\pi + \delta$ |

*Note*: The photon wavelength is $\lambda$ and the mirror location at u is adjustable using a piezo-driven mechanical system to vary the path difference $\delta$. Here, BS is a beam splitter, M is a mirror, the numbers refer to the Principles listed above, and the numbers in brackets are the phase shifts.[16]

---

[16]The beam splitter $BS_2$ is coated on the right side in Figure 3.3 which is why the u path takes two internal paths before reflection to $MZ_2$ (with no phase shift), while the d path reflects to $MZ_1$ without passing into the beam splitter (with $\pi$ phase shift). A path inside a beam splitter is denoted by BS5 in Table 3.3.

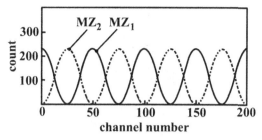

Figure 3.4   Superimposed curves showing the number of counts in outputs $MZ_1$ and $MZ_2$ as a function of the path difference $\delta$. Starting at $\delta = 0$, one channel corresponds to a $\lambda/50$ variation of $\delta$. Total counting time 15 s. Adapted from Grangier *et al.* (1986; Figure 4, 178).

authors remark that "we are compelled to use a wave picture (the electromagnetic field is coherently split on a beam splitter) to interpret the second (interference) experiment" (Grangier *et al.* 1986, 178–9).

### 3.4.2   *Proposed photon model*

This brings us to our proposed model for a single photon. In the first Grangier experiment, the consideration of an arbitrarily extended transverse structure to the beam has small relevance since no opportunity is afforded for superposition of diverse parts of the single-photon front. However, for the second experiment, we now introduce the model of an array of advancing helical space dislocations (HSDs) or screw threads; see Figure 3.5.[17] The front is divided both at $BS_1$ and $BS_2$ with each individual HSD thread on the transverse array carrying the possibility of exciting the detector it impinges on at $MZ_1$ and $MZ_2$. The process of absorption at a detector must then involve the participation of a single HSD in the array recombining and interfering with another HSD on the array,[18] during which a

---

[17]This is also a transversely iterated circular polarization model.

[18]The circular area for a HSD has diameter $\lambda/\pi$.

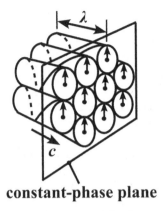

**constant-phase plane**

Figure 3.5   Alternative single-photon model of a transverse array of HSD. A constant-phase front of a right-circular polarization single photon, consisting of a transversely iterated array of left-wound HSD advancing with speed $c$ through a stationary plane. Each single HSD has longitudinal extent $\lambda$, linear action $h$, and a cross-sectional area $\lambda^2/4\pi$. The arrows indicate the phase.

single quantum of linear action $h$ and a spin angular momentum of magnitude $\hbar$ are transferred.[19] Let us consider this in more detail.

Consider the Mach–Zender interferometer in Figure 3.3 in relation to Figure 3.5. A single circular polarization HSD at A on the transverse front, prior to $BS_1$, is to be absorbed at $MZ_1$ as a photon registration. If there is interference, there must be a similar HSD at B, another part of the array, separated laterally from A that takes the alternative path and also arrives at the detector $MZ_1$ to coincide and interfere with A. By considering their phase relation, the HSD at B will either enhance the registration of A or diminish it depending on the path difference. Let us consider that A takes path d, and B takes path u. With a channel increment of $\lambda/50$, where $\lambda$ is the wavelength of A, if a complete annihilation occurs for channel $n$, then

---

[19] A linear polarization HSD can only transfer linear momentum. However, an absorbed circular polarization photon is a single-turn helix which, when moving at speed $c$, generates both linear and spin angular momentum in a plane perpendicular to the direction of the photon. Since the action is quantized, the linear momentum is $= h/\lambda$, while the spin angular momentum has magnitude $h/2\pi = pa$, where $\lambda = 2\pi a$, and $a$ is the helix cross-section radius, see Poynting (1909, 560–67).

we see from Figure 3.4 that it also occurs for channel $n + 50$. So, B maintains the same wavelength as A, reflects and transmits according to the same rules as A, and is also capable of effecting a registration like A. However, there is a further consideration. If interference still occurs for changes in path difference greater than one wavelength, each tube on the front must also be iterated longitudinally, that is, the extent of it is a whole number $n > 1$ wavelengths.

In summary, we now have two different interpretations as to how a coincidence registration might occur in the beam splitter section of the apparatus.

(1) **Traditional quantum mechanics hypothesis:** Different SPDC events allow more than one localized single photon to pass through the beam splitter during the gate with each capable of registering only once (Thorn *et al.* 2004). Fortuitously, the detectors are triggered by two separate localized photons. This preserves the idea that if a coincidence registration occurs, it cannot be due to a single photon.

(2) **Proposed HSD-array model:** There is a single-photon front passing through the beam splitter during a gate that takes the form of a transverse array of HSD, each tube on the front having the potential to effect a registration. Given that an SPDS event occurs, which is known from an idler photon detection, when one of the corresponding signal HSD in the array excites a detector, $MZ_1$ say, the rest of the divided front might effect a further registration at the other detector $MZ_2$ with a probability similar to those given in Tables 3.1 and 3.2. Here, a coincidence registration arises from a splitting of the transverse array of HSD.[20] When the parts of the divided front are recombined in an interferometer, two or more HSD from the front have an opportunity to overlap both in space and time, and thereby interfere.

---

[20] At the 1927 Solvay Conference, in an argument for incompleteness, Einstein suggested without further analysis that the wave function $\psi$ in quantum mechanics might represent 'an ensemble (or cloud) of electrons' (Bacciagaluppi and Valentini 2009, 194).

The term 'single photon' seems not to do justice to this model, for our assumption is that an atomic transition emits an array of HSD arranged laterally, which might also be longitudinally iterated in increments of one wavelength with phase-matched ends. Also, it appears from the Grangier experiment that the limits that determine the conditions for absorption are met only rarely. It is supposed that the conditions for excitation of a detector must be related to the orientation and state of the electron orbit in the atomic target.

## 3.5  Temporal Second-Order Correlation Function

A calculation can be carried out that quantifies the extent of the quantum or classical behavior of photons. The first practical use of a spatial second-order correlation function, involving quantities quadratic in the field strengths, was made by Hanbury Brown and Twiss (1956a, 1956b). They pointed two telescopes a known distance $D$ apart at the brightest star Sirius and found that the collected intensities, filtered to be monochromatic, were so correlated as to allow the angular diameter of the star to be obtained. The light from the left extremity of the star into both telescopes produces one intensity profile, while that from the right extremity produces a similar intensity profile whose center is displaced from the first. The distance $D$ between the telescopes is increased until the first-order fringes disappear at which point $\varphi \approx d/L$, where $d$ is the star diameter and $L$ is the stellar distance.[21] Here, angle $\varphi$ is that between the rays into one of the detectors from the star extremities.

We now return to the coincidence-counting experiments of Section 3.2 in which SPDC was used to produce an idler and a signal photon. The detection of the idler photon meant that the signal photon was known to exist and was advancing on the 50–50 beam splitter. However, the values of $P_C$ recorded in Table 3.1 suggest that there is a definite probability much less than unity that the signal photon can excite a detector given that a photon is incident on it. Let us reconsider the case of a beam

---

[21]The small angle approximation $\varphi \approx \tan \varphi$ is used.

Table 3.4   A quantum second-order correlation in which one of the detectors but not both registers a count.

| $i$ | 1 | 2 | 3 | 4 | 5 | 6 | 7 | 8 | 9 | 10 | 11 | 12 |
|---|---|---|---|---|---|---|---|---|---|---|---|---|
| $I_1(t_i)$ | 1 | 0 | 0 | 1 | 1 | 1 | 0 | 1 | 0 | 0 | 0 | 0 |
| $I_2(t_i)$ | 0 | 1 | 1 | 0 | 0 | 0 | 1 | 0 | 1 | 1 | 1 | 1 |

*Note*: Registrations recorded at detectors 1 and 2 are taken over a time interval divided into 12 equal increments. A registration is indicated by a 1 and a non-registration by a 0.

splitter placed in the path of a signal photon $v_2$; see Figure 3.2. Suppose that a signal-photon registration at either one of the detectors $PM_r$ and $PM_t$ at the beam splitter ports initiates a countable time increment or gate. Let us label these detectors 1 and 2. The sum of such increments is to constitute the measurement time interval. The registration of a photon at detector 1 is indicated by $I_1 = 1$ and its non-detection by $I_1 = 0$. Similarly, for detector 2, we have by $I_2 = 1$ and $I_2 = 0$. First, we examine a quantum model where one of the detectors, but not both, triggers a time increment (or gate). An example is shown in Table 3.4 for $n = 12$ time increments.

The second-order temporal correlation function is defined as follows[22]:

$$g^{(2)}(\tau) = \frac{\langle I_1(t)I_2(t+\tau)\rangle}{\langle I_1(t)\rangle\langle I_2(t)\rangle} \tag{3.2}$$

If the measurements are taken simultaneously, $\tau = 0$, then we have

$$g^{(2)}(0) = \frac{\langle I_1(t)I_2(t)\rangle}{\langle I_1(t)\rangle\langle I_2(t)\rangle} \tag{3.3}$$

Here, the angled brackets indicate a mean value summed over a time interval.

---

[22] A mathematical treatment of coherence functions has been given by Glauber (1963).

In Table 3.4, we consider a quantum model and if we now calculate the second-order correlation function in (3.3), we arrive at

$$g_q^{(2)}(0) = \frac{\langle I_1(t) I_2(t) \rangle}{\langle I_1(t) \rangle \langle I_2(t) \rangle} = \frac{\frac{0}{12}}{\left(\frac{5}{12}\right)\left(\frac{7}{12}\right)} = 0 \qquad (3.4)$$

For a classical wave model, a photon front incident on a detector necessarily produces a registration and then Table 3.4 would contain only ones and no zeroes. The calculation for (3.3) would then become

$$g_c^{(2)}(0) = \frac{\langle I_1(t) I_2(t) \rangle}{\langle I_1(t) \rangle \langle I_2(t) \rangle} = \frac{\frac{12}{12}}{\left(\frac{12}{12}\right)\left(\frac{12}{12}\right)} = 1 \qquad (3.5)$$

However, it is suggested here that neither the quantum nor the classical model gives the 'real' state of affairs. Instead, Table 3.1 records experiments that suggest the following. Given that a single photon is incident on the signal detector (known from detection of the idler), there is a probability of excitation for the signal detector that is less than one, that is,

$$7.494 \times 10^{-4} \le P_C \le 1.03 \times 10^{-2} \qquad (3.6)$$

In other words, this is the probability of a coincident detection. We now construct Table 3.5 where in addition to only one detector being excited, occasional coincidences occur. When this occurs, a '1' appears twice in the column of a time increment. Ignoring (3.6), let us by way of example arbitrarily set the probability of coincidence to be $P_C = 1/3$ so that with $n = 12$ time increments there are $n/3 = 4$ coincidences.

The calculation for (3.3) is now

$$g_c^{(2)}(0) = \frac{\langle I_1(t) I_2(t) \rangle}{\langle I_1(t) \rangle \langle I_2(t) \rangle} = \frac{\frac{4}{12}}{\left(\frac{7}{12}\right)\left(\frac{9}{12}\right)} = \frac{16}{21} < 1 \qquad (3.7)$$

This brings out the essence of 'real' quantum mechanical behavior.[23] The reason there is interference when the beam splitter outputs are recombined for a single photon (see Section 3.4) is because the photon has 'real' transverse iteration. So, we are entitled to ask

---

[23]Thorn *et al.* (2004) give $g^{(2)}(0) = 0.0177 \pm 0.0026$.

Table 3.5   A classical-quantum second-order correlation in which only one or both detectors produce a registration.

| $i$ | 1 | 2 | 3 | 4 | 5 | 6 | 7 | 8 | 9 | 10 | 11 | 12 |
|---|---|---|---|---|---|---|---|---|---|---|---|---|
| $I_1(t_i)$ | 1 | 0 | 0 | 1 | 1 | 1 | 0 | 1 | 1 | 0 | 1 | 0 |
| $I_2(t_i)$ | 0 | 1 | 1 | 1 | 0 | 0 | 1 | 1 | 1 | 1 | 1 | 1 |

*Note*: Registrations recorded at detectors 1 and 2 are taken over a time interval divided into 12 equal increments. A registration is indicated by a 1 and a non-registration by a 0. The probability of registration given that a photon is incident is arbitrarily taken to be 1/3.

the following: Why is it that the detections after recombination are not all coincidences? For a two-slit interference experiment with electrons, there is good reason why Feynman (2006, III.1-5) formed the view that "each electron either goes through hole 1 or it goes through hole 2 [but not both]".[24] When he made this assertion, the sensitivity of the experimental apparatus was such that the probability of both electrons being detected as a coincidence was too low to be noticeable. This is what (3.6) testifies. Of course, one could claim that the coincidences are due to spurious photons from other SPDC events entering the system, but it is an explanation that keeps us locked in difficulty.[25] The advantage of the interpretation introduced here is that the conjecture that each HSD on the photon front is capable of exciting a detector gives us a way out of our quantum paradox. A single photon *does* go through both slits. It is just that if one part of the division registers with one detector, the probability that the other registers with another detector can be as

---

[24]With a light source behind the slit holes to scatter the electron, he also says "we *also see* a flash of light *either* near hole 1 *or* near hole 2 but *never* both at once!" (Feynman 2006, III.1.7).

[25]Colloidal quantum dot technology has been used to produce a "highly suppressed multi-photon-emission probability", that is, photons that are highly likely to arrive one by one (i.e., they are 'antibunched'). Coincidence counts obtained from the exit ports of a beam splitter still occur in this quasi-pure single-photon environment; see Lin *et al.* (2017, Figs. 1e and 4).

low as that given in (3.6), so the coincidence although present might not be registered.

## 3.6   Local or Non-local Interference?

A two-photon experiment that sets out to show that a photon in glass travels at the group velocity far from any resonances has been reported by Steinberg *et al.* (1992). As they remark, "large bandwidths are necessary for high time resolution, but in dispersive media they also lead to large wave-packet spreading, and hence to a reduction in resolution" (Steinberg *et al.* 1992, 2421). However, the expected dispersion associated with a wave packet, in which no two of its composite frequencies has the same transit time through the medium, is not observed in their experiment. They attribute this "peculiar cancellation effect" (Steinberg *et al.* 1992, 2424) — by association not explanation — to a "non-local purely quantum mechanical effect" (Steinberg *et al.* 1992, 2421).

Our aim here is to analyze this experiment and to show that not only is non-local interference not a 'real' interference effect but that this mysterious 'cancellation' is evidence that a photon is *not* a wave packet (nor a pulse).[26] Instead, it is shown how the notion of non-local interference can be replaced by a classical analysis using the visualizable HSD-array single-photon model.

The apparatus shown in Figure 3.6 passes an argon ion laser beam at 351 nm through a potassium diphosphate (KDP) crystal to create spontaneous down conversion (SPD) which, with the assistance of irises I1 and I2 near the detectors D1 and D2, produces nearly degenerate pairs centered at 702 nm. The photons are Type I anticorrelated and exit the crystal on opposite sides of a cone whose axis is that of the UV pump.[27] The signal photon is directed through

---

[26]Those physicists who have abandoned the idea that an interaction between two substances requires both spatial and temporal coincidence are not attempting to understand Nature. Their pernicious constructions only have the effect of deflecting researchers away from serious attempts at finding 'real' solutions.

[27]Each photon is 'anticorrelated' means that after the beam splitter, a photon should be able to travel along only one path rather than subdivide. The sum of the frequencies (energies) of the two 'anticorrelated' photons totals that of the pump

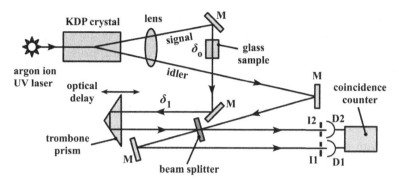

Figure 3.6    Measurement of the photon propagation time delay due to the introduction of a glass sample. The M indicate mirrors, D1 and D2 show detectors, and the photons impinge on a 50/50 beam splitter. The trombone prism can be translated to introduce a time delay into the signal path (Steinberg *et al.* 1992, Fig. 1).

the glass sample and can be delayed by translation of the trombone prism, while the idler travels through air. The photons are brought into proximity at a symmetrical beam splitter which leads to the two detectors wired to a coincidence counter, the whole set-up being a variation of the Hong–Ou–Mandel interferometer (Hong *et al.* 1987). The detectors are EGG C30909S silicon avalanche photodiodes positioned at the two output ports of the beam splitter. When one is excited, a coincidence is counted if the second detector fires within 5 ns of the first. When the glass sample is introduced into the path of the signal photon, the effect on the coincidence output can be observed. By varying the optical delay $\delta_1$ to restore the position of the coincidence minimum, a measurement of the delay time caused by the glass on the signal photon can be made. Some variation of the trombone prism position without the glass is needed at the start to calibrate the initial coincidence null to a zero delay.

### 3.6.1    *Quantum mechanical analysis*

Let the signal photon be 'S', the idler 'I', detector 1 denoted by '1', and detector 2 by '2'. The experiment assumes that the photon is a

---

laser, so if one frequency could be made to increase, the other would decrease in a linear relationship (negative gradient or 'anti'-correlation).

localized entity (anticorrelated), in this case a wave packet, that is, indivisible at the beam splitter. The concept demands a significant bandwidth of frequency components in order to represent it. There are two ways that a coincidence can occur at the detectors: both photons are reflected at the beam splitter (r–r) as S1–I2; or both are transmitted (t–t) as S2–I1. The two photons travel to different detectors in each case, and there is no division of a single photon at the beam splitter. As far as the coincidence event is concerned, there is no observational way of knowing which of the two combinations produces it, and so they become indistinguishable alternatives. The rules for combining quantum-mechanical amplitudes have been given by Richard Feynman:

> Provided that the two particles do not interact, the amplitude that one particle will do one thing *and* the other one something else is the product of the two amplitudes that the two particles would do the two things separately. (Feynman 2006, III.3-4)

This echoes the probability rule for independent events but instead employs amplitudes. Its counterintuitive character gives it the appearance of a definition. Also,

> When a particle [or combination of them] can reach a given state by two possible routes, the total amplitude for the process is the *sum of the amplitudes* for the two routes considered separately. (Feynman 2006, III.3-3)

The amplitudes are complex numbers and since each reflection at the beam splitter produces a $\pi/2$ phase shift with respect to the transmitted component (Steinberg *et al.* 1992, 2422), then the total amplitude for 'S1–I2 (r–r) or S2–I1 (t–t)' can be written as

$$\phi = \frac{e^{i\pi/2}}{\sqrt{2}} \cdot \frac{e^{i\pi/2}}{\sqrt{2}} + \frac{e^{i0}}{\sqrt{2}} \cdot \frac{e^{i0}}{\sqrt{2}} = \frac{i}{\sqrt{2}} \cdot \frac{i}{\sqrt{2}} + \frac{1}{\sqrt{2}} \cdot \frac{1}{\sqrt{2}} = 0 \quad (3.8)$$

Since the probability of a photon arriving at exactly one of the two detectors is $1/2$, each component (e.g., S1) in (3.8) has the square of its amplitude set to this value. The probability of a coincidence is the square of $\phi$. The zero result can be interpreted as a *non-local* destructive interference between the two alternatives by which a coincidence count can occur.

Let us take some time out to consider what all this means. In classical wave theory, we only add amplitudes for substances that are spatially and temporally *coincident*. The wave has substance and this is the nature of 'real' interference. In the present quantum mechanical treatment, the amplitudes are added for temporal but not spatial coincidence. However, this is not an interference between amplitudes of 'real' substances. This is an interference between logical possibilities. In fact, wave packets have such a small lateral extent that even if those representing the two different photons managed to fall simultaneously on the same detector, they would be unlikely to interact. There would be negligible interference of substances. So, when one asserts that logical possibilities at *different* detectors are allowed to interfere, one wonders if one is being asked to believe that there is a supernatural interaction between *substances* that arrive simultaneously at different detectors.

Now, Steinberg *et al.* (1992, 2422) claim that the zero result from the interference of these logical possibilities (S1–I2 and S2–I1) implies that there is a coincidence null, and that this informs us of the 'real' situation that "the two photons exit the same port of the beam splitter" into the same detector as a singles count. There is not even a demand for spatial coincidence of the wave packets in a detector. It is enough that the two photons take the same path out of the beam splitter simultaneously. Now, there are two difficulties here. The first is that the height of the coincidence null in Figure 3.7(a) is about 85% of the maximum coincidence count. So, there is still a significant coincidence count even though the 'logical' result is zero. Why has this purely 'logical' destructive interference not eliminated *all* coincidences? The second problem is as follows. By introducing phase shifts at the beam splitter, it is clear that a wave theory has been invoked. However, this wave is only a computational device. It has no substance whatsoever, yet curiously it can be phase shifted by the material constituting the beam splitter. One wonders how a substance manages to causally connect to a non-substance.[28] So, in the preceding paragraph, it is logical possibilities or non-substances that are interacting, while in this paragraph we find an interaction between a non-substance and a substance. It is not an alternative explanation that we need to find, because in truth we are not in possession of one to find an alternative to.[29] We just need to find an explanation.

So far, the analysis has been applied to two photons arriving at the beam splitter at the same moment. However, a time delay can be introduced into the signal path by translating the trombone prism and/or introducing the glass, and then we have a separation of the centers of the S1 and I2 wave packets on the time axis, as well as those for S2 and I1. The destructive interference and the null coincidence result then begin to dissolve. The authors remark that "the two wave packets no longer overlap in time, and the two Feynman paths become, in principle, distinguishable" (Steinberg *et al.*, 1992, 2442). The two photons are now assumed to exit different ports of the beam splitter for a coincidence count. The detector time resolution, being about 100 ps, is large enough to register the two photons as coincident even though they are slightly displaced in time.

### 3.6.2   *Classical transverse HSD-array analysis*

In the new classical HSD–array model presented here, a single photon is now assumed to have 'real' transverse extent (Figure 3.5) so it can divide at the beam splitter, with components from each photon interacting in the same detector; see Figure 3.6.[30] The claim here is that Dirac was incorrect in asserting that a photon can only interact with itself. For two photons identical in frequency, polarization, and space–time location, interference *can* occur. The relative phase

---

[28] Again, is one being asked to believe that the supernatural has intruded on the natural world?

[29] The Emperor has no clothes, but the problem is not so much a lack of awareness, but that no one can find clothes to fit him.

[30] This is in opposition to the quantum mechanical model of two localized photons, which strike the detector simultaneously at different locations.

Table 3.6  Relative phase shifts of arrivals at detectors D1 and D2 for a symmetric beam splitter, calculated from the difference of phase totals.

| Destination | Signal (S) | Idler (I) | S − I relative phase shift $\Delta$ |
|---|---|---|---|
| Detector D1 | M4($\pi$), $\delta_o$, M4($\pi$), $\delta_1$, T5(0), T3(0), T5(0), T3(0), T5(0), BS6($\pi/2$), M4($\pi$) (reflect) | M4($\pi$), M4($\pi$) (transmit) | $3\pi/2 + \delta$ |
| Detector D2 | M4($\pi$), $\delta_o$, M4($\pi$), $\delta_1$, T5(0), T3(0), T5(0), T3(0), T5(0) (transmit) | M4($\pi$), BS6($\pi/2$) (reflect) | $\pi/2 + \delta$ |

*Note*: The phase shifts are given in brackets. The resultant phase shift $\delta = \delta_o + \delta_1$.

shift in a particular detector between the components from the two photons is given in Table 3.6, column 4. The second and third columns of Table 3.6 show the phase shifts experienced by the signal and idler rays on their way to detectors D1 and D2. The letters refer to an object in Figure 3.6, while the numbers refer to the *Principles of reflection and transmission* in Section 3.4.1. The phase shift for each object is in brackets. The trombone prism T produces no phase shift in the signal photon from reflection since both internal reflections have a less dense medium (air) behind the transmission medium (in Table 3.6, T3 arises from trombone T and *Principle* 3). In fact, Steinberg *et al.* (1992) assign no effect to either reflection (T3) or transmission (T5) through the trombone prism. Each mirror M leaves a $\pi$ phase shift (M4). A sixth *Principle* can be added to create the $\pi/2$ phase shift referred to as BS6 in Table 3.6 for the beam splitter. The beam splitter BS is symmetric so that the same phase shift $\pi/2$ occurs on each side after reflection.[31] Transmission through the beam splitter has no effect on the phase. Let the glass

---

[31]The sum of the reflection phase shifts minus the transmission phase shifts must always be $\pi$ to conserve energy, see for example (Kučera 2007, 4). Also, "the reflection amplitude of a lossless beam splitter is $\pm\pi/2$ out of phase with respect to its transmission amplitude" (Steinberg *et al.* 1992, 2422).

sample produce a phase shift of $\delta_o$. A translation of the trombone prism allows the signal to differ in spatial displacement from the idler path and this also produces a phase shift $\delta_1$. The final column in Table 3.6 is the difference between the total phase shift in the signal and idler paths $(S - I)$, where $\delta = \delta_o + \delta_1$.

So how does our 'realistic interference' interpretation account for the 85% (non-zero) coincidence rate at the incomplete null in Figure 3.7(a)? Let us reiterate our assumption that HSD tubes from both the idler and signal HSD fronts enter each detector whatever the value of $\delta$. If they do not coincide both spatially and temporally in a detector then there is no possibility of interference, and so there is no prohibition on detector registrations. In that case, we obtain the maximum coincidence rate of approximately 1200/s each side of the null.

Those tubes that coincide spatially could produce a destructive interference and reduce the coincidence count if there was a degree of temporal coincidence. A null could occur at $\delta = -\pi/2$, giving the relative phase shift $\Delta$ in Table 3.6 as $\pi$ at detector D1 (destructive) and zero at D2 (constructive). There is definitely a reduced coincidence count here. A relaxation of the null could occur each side of this. At $\delta = -\pi$, we have $\Delta$ as $\pi/2$ at D1 (partial destructive) and $-\pi/2$ at D2 (partial destructive). At $\delta = 0$, we find $\Delta$ to be $3\pi/2$ at D1 (partial destructive) and $\pi/2$ at D2 (partial destructive). Any further shift in $\delta$ away from $\delta = -\pi/2$ each side of it produces an insignificant temporal coincidence and a high coincidence rate.

### 3.6.3   *Non-dispersion of the photon*

Figure 3.7(a) depicts the coincidence null before the glass is introduced, while Figure 3.7(b) shows it after. In the latter, the null has shifted by 35,219 fs indicating that passing the signal photon through glass has introduced a time delay corresponding to $\delta_o$, shifting it along the time axis with respect to the idler photon. To restore the interference in the detector, the trombone prism must

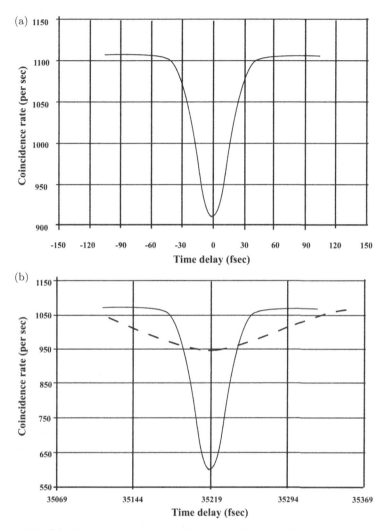

Figure 3.7   (a) Coincidence rate as a function of the relative optical time delay in the interferometer. The solid line is a Gaussian fit, with an RMS width of 15 fs. For Steinberg *et al.* (1992), the profile serves as a map of the overlapping photon wave packets. (b) Coincidence profile after a 1/2″ piece of SF11 glass is inserted in the signal arm of the interferometer. The dip is now shifted by 35219 fs, but no broadening is observed. Classically, a 15-fs pulse would broaden to at least 60 fs, shown for comparison as a dashed line (Steinberg *et al.* 1992, Fig. 3).

be given a compensating translation. Steinberg *et al.* (1992, 2424) remark on "a peculiar cancellation effect which for most intents and purposes negates the effect of group velocity dispersion" in the glass. The dotted line in Figure 3.7(b) shows what the curve would have looked like had dispersion occurred: "a 15 fs pulse would broaden to at least 60 fs"." It does not broaden, and it is suggested here that the reason is because a photon has no dispersion. The photon is neither a pulse nor a wave packet.

According to the classical account given here, a photon must have transverse extent if there is to be any 'real' interference between components from the two photons in a detector. The conclusion is that Dirac (2000, 9) had insufficient facts available to him when he asserted that "interference between two photons never occurs".

## 3.7  Concluding Remarks

If we return to the experiment of Grangier *et al.* (1986) in Section 3.4 and examine a single photon passing through a Mach–Zender interferometer, the conclusion seems unavoidable that in order for superimposed parts of the front to be capable of constructive and destructive interference, there must be 'real' regions of identical phase and structure on the beam front. Since circular polarization (HSD) tubes serve well as a foundation for a theory of circularly polarized light rays, it is suggested here that a single-photon beam is constructed from an array of such tubes distributed uniformly over an advancing front; see Figure 3.5.

The bounded areas on the beam front are tube cross-sections, each guiding the passage of an identical non-rotating screw thread at light speed c, and inducing a circularly polarized rotation in a stationary plane (set perpendicular to the optic axis) through which the screw thread passes. This undoubtedly leads to an unlimited energy content taken across an unbounded single-photon front, but it should be kept in view that access to the energy of any tube on the front is only possible on the infrequent occasion when the relative orientation and phase of the target electron in the detection surface are favorable. We have already seen from the work of Thorn *et al.*

(2004) that the coincidence probability might be as low as $P_c = 3.54 \times 10^{-4}$. Nevertheless, if a material can be found and its condition modified to present favorable electron orientations to an impinging single-photon array, then this might allow an amplification of the energy that can presently be extracted from a single-photon front.[32]

The Grangier experiment has brought out two seemingly irreconcilable aspects of single-photon behavior. Feynman gives the mathematical relations as follows[33]:

(1) The probability of an event in an ideal experiment is given by the square of the absolute value of a complex number $\phi$ which is called the probability amplitude:

$$P = \text{probability}$$
$$\phi = \text{probability amplitude} \quad (3.9)$$
$$P = \phi^2$$

(2) If an experiment is performed which is capable of determining whether one or another alternative is actually taken, the probability of the event is the sum of the probabilities for each alternative. The interference is lost:

$$P = P_1 + P_2 \quad (3.10)$$

(3) When an event can occur in several alternative ways, the probability amplitude for the event is the sum of the probability amplitudes for each way considered separately. There is interference:

$$\phi = \phi_1 + \phi_2$$
$$P = |\phi|^2 \quad (3.11)$$

---

[32] An electron in the target surface viewed as a ring would need to have its normal vector approximately aligned with the incoming SAM vector. It is conceivable that a material might be employed that facilitates the alignment of the surface electron rings, perhaps under the influence of an appropriately directed magnetic field.

[33] See Equations (1.6)–(1.8) in Feynman *et al.* (2006, III.1–10).

Equation (3.10) applies to the first part of the Grangier experiment in which the front is divided with a beam splitter. In relation to the HSD array model suggested here, Feynman implies that there is only one single localized area on the front capable of triggering a detector and so it must take either one path or the other at a beam splitter but not both. However, as we have seen, a small coincidence count *does* occur. Equation (3.11) applies to the second part of the Grangier experiment in which an interferometer is installed in place of the beam splitter. This now provides an opportunity for the two paths to interact and interfere. The notion that the single-photon front is an array of tubes sits easier with this case than with the previous one. Here, we fully expect a substance to be present on both paths in the interferometer if a 'real' local interference is to take place. The suggestion here is that a single photon does not present a *single* opportunity for registration but an *array* of opportunities which are distributed between the two paths. A coincidence occurs so infrequently that to all appearances only one occurs at a time. However, in principle, *any number of registrations* can occur simultaneously. This only becomes evident in experiments involving a large number of photons.

Feynman (2006, III.1–10) has remarked that "No one has found any machinery behind the law [of quantum mechanical probability amplitudes...] We have no ideas about a more basic mechanism from which these results can be deduced." However, it is suggested here that the difficulties that have plagued the interpretation of quantum mechanics persist only so long as the assumption is upheld that only a single localized area of the photon front can affect a detector at any moment. Once we remove this prohibition, and admit a coexisting array of real localized areas on the front, each capable of affecting a detector, then the phenomenon of interference becomes an interaction between *real* coexisting substances instead of one between *logical* coexistences.

## References

Ahmed, M., A. Amponsah, A. Emmanuel, and H. Issake. 'Source of photon pairs using spontaneous parametric down-conversion process'. *International Journal of Innovation and Applied Studies*, 9 (2014): 734–743.

Aspect, A., P. Grangier, and G. Roger. 'Experimental tests of realistic local theories via Bell's theorem'.*Physical Review Letters*, 47 (1981): 460–463.

Bacciagaluppi, G., and A. Valentini. *Quantum Theory at the Crossroads. Reconsidering the 1927 Solvay Conference*. Cambridge University Press, 2019.

Dempster, A. J., and H. F. Batho. 'Light quanta and interference'. *Physical Review*, 30 (1927): 644–648.

Dirac, P. *The Principles of Quantum Mechanics*, Fourth edition. Oxford: Clarendon Press, 2000.

Dontsov, P. Y., and A. I. Baz. 'Interference experiments with statistically independent photons'. *Soviet Physics — Journal of Experimental and Theoretical Physics (JETP: English translation)*, 25 (1967): 1–5.

Feynman, R., R. B. Leighton, and M. Sands. *The Feynman Lectures on Physics. The Definitive Edition*, Fourth edition, 3 vols. Pearson, 2006.

Glauber, R. J. 'The quantum theory of optical coherence'. *Physical Review*, 130, 6 (1963): 2529–2539.

Grangier, P., G. R., and A. Aspect. 'Experimental evidence for a photon anticorrelation effect on a beam splitter: a new light on single-photon interferences'. *Europhysics Letters*, 1 (1986): 173179.

Hanbury Brown, R., and R. Q. Twiss. 'Correlations between photons in two coherent beams of light'. *Nature*, 177 (1956a): 27–29.

Hanbury Brown, R., and R. Q. Twiss. 'A test of a new type of stellar interferometer on Sirius'. *Nature*, 178 (1956b): 1046–1048.

Hong C. K., Z. Y. Ou, and L. Mandel. 'Measurement of subpicosecond time intervals between two photons by interference'. *Physics Review Letters*, 59 (1987): 2044–2046.

Kučera, P. 'Quantum description of optical devices used in interferometry'. *Radioengineering*, 16(3) (2007): ISSN 1210–2512.

Kwiat, P. G., E. Waks, A. G. White, I. Appelbaum, and P. H. Eberhard. 'Ultra-bright source of polarization-entangled photons'. *Physical Review A*, 60 (1999): 773–776.

Lin, X., X. Dai, C. Pu, Y. Deng, Y. Niu, L. Tong, W. Fang, Y. Jin, and X. Peng. 'Electrically driven single-photon sources based on colloidal quantum dots with near-optimal antibunching at room temperature'. *Nature Communications*, 8 (2017): Article number 1132.

Noh, T.-G., H. Kim, C. Ju Youn, S.-B. Cho, J. Hong, and T. Zyung. 'Nonlinear correlated photon pair source in the 1550nm telecommunication band'. *Optics Express*, 14, 7 (2006): 2805–2810.

Pipkin, F. M. 'Atomic physics tests of the basic concepts in quantum mechanics'. In D. R. Bates and Benjamin Bedereson (eds.), *Advances in Atomic and Molecular Physics*, Vol. 14. Academic Press, 1979, pp. 281–340.

Reynolds, G. T., K. Spartalian, and D. B. Scarl. 'Interference effects produced by single photons'. *NuovoCimento*, 61 B (1969): 355–364.

Steinberg, A. M., P. G. Kwait, and R. Y. Chiao. 'Dispersion cancellation in a measurement of the single-photon propagation velocity in glass'. *Physical Review Letters*, 68 (1992): 2421–2424.

Taylor, G. I. 'Interference fringes with feeble light'. *Proceedings of the Cambridge Philosophical Society*, 15 (1909): 114–115.

Thorn, J., M. S. Neel, V. W. Donato, G. S. Green, R. E. Davies, and M. Beck. 'Observing the quantum behaviour of light in an undergraduate laboratory'. *American Journal of Physics*, 72 (2004): 1210–1219.

Tonomura, A., J. Endo, T. Matsuda, T. Kawasaki, and H. Ezawa. 'Demonstration of single-electron build up of an interference pattern'. *American Journal of Physics*, 57 (1989): 117–120.

## Chapter 4

# A Longitudinally Iterated Single Photon

*A review of the Bose counting method for indistinguishable photons is given to derive Planck's radiation law. A new counting method is introduced based on the notion of rotating sequences of photons and cells to obtain the Bose results from the assumption of distinguishable photons. Here, cells are distinguished by their direction of approach to an area element in the surface of a blackbody cavity. This shows that it is not necessary to employ concepts that correspond only to observables in order to obtain the Planck law.*

## 4.1 A Brief History

One application of the helical space dislocation (HSD) model is to the theory of blackbody radiation and this will now be developed here.[1] However, a historical overview will be instructive. The discovery that Planck (1900) made at the beginning of the twentieth century was a formula for the energy density in the frequency interval $(\nu, \nu + d\nu)$ that matched the data presented to him by Rubens and Kurlbaum (1901) in October of that year,

$$u_\nu d\nu = \frac{A\nu^3}{(e^{B\nu/T} - 1)} d\nu \tag{4.1}$$

Despite its empirical success, its conceptual basis was deeply flawed.

(1) In its attempt to use Boltzmann's statistical counting method, an ill-defined population was given for the probability calculation (Jeans 1905, 293).

---

[1] A history of the subject is set out in detail in Clarke (2017, Chapter 4).

(2) An oscillator of a given frequency was confined to exchanges with oscillators of the same frequency, and no mechanism was available for redistributing its energy among other modes (Kuhn 1978, 166). Planck eventually acknowledged this defect six years later.[2]

(3) There was an internal contradiction in the conceptual basis: the continuous field theory of Maxwell in obtaining a relation between the energy density and the average energy per oscillator; and the counting of discrete units of energy using modified Boltzmann statistics (Einstein 1905, 132).

Einstein (1916) partly overcame these difficulties by deriving the law from considerations of emission and absorption rates. The conflict inherent in (3) was removed, while the notion that a resonator with frequency $\nu$ can only take on the energy values $\varepsilon_n = nh\nu$, where $n = 0, 1, 2, \ldots$ was retained. To effect the derivation, Einstein identified three emission and absorption mechanisms as follows:

(1) The probability $dW$ that an emission of energy $\varepsilon_m - \varepsilon_n$ ($m > n$) with frequency $\nu$ takes place in an arbitarily small time $dt$ independent of external causes is

$$dW = A_m^n dt \qquad (4.2)$$

(2) For an induced emission, the resonator changes state from $m$ to $n$ ($m > n$) in virtue of incident radiation with density $\rho$, and the probability of doing so is

$$dW = B_m^n \rho dt \qquad (4.3)$$

(3) Absorption from incident radiation with density $\rho$ has the probability

$$dW = B_n^m \rho dt \qquad (4.4)$$

---

[2] "The oscillators which provide the basis for the present treatment influence only the intensities of the radiation corresponding to their own natural frequencies" (Kuhn 1978, 161).

Now, it is undoubtedly the case that induced emissions can be obtained in lasers, but this does not necessarily mean that the foundations of Einstein's (1916) theory are correct. Its weakness lies in Einstein's admission that the energy levels involved in the transitions are *internal energies*.[3] The reason that this cannot be the case has been illustrated by Sir Neville Mott who has calculated that for a monatomic gas with $kT \sim 1/40$ eV (room temperature), an energy of $\sim 5$ eV must be supplied to remove the electron from its ground state and raise it to its first excited state (Mott 1964). For hydrogen, the energy of this particular transition is $\sim 10.2$ eV. Consider a gas of atomic hydrogen that has been set up in thermal equilibrium with heat radiation in a perfectly reflecting container. Let us also select a temperature for the gas such that $kT \ll 10.2$ eV. Under these conditions, the rate of transitions between internal electron energy levels is insufficient to maintain equilibrium. So, our conclusion must be that the theory of induced emissions cannot accommodate heat radiation.[4]

## 4.2 The Bose Counting Method

The work of Satyendra Nath Bose on Planck's law in 1924, which was sanctioned by Einstein, also deserves analysis (Bose 1924).[5] There are two parts to his derivation:

(1) for the frequency interval $(\nu, \nu + d\nu)$, an expression for the number of cells $A$ in a 6-D phase space with volume $h^3$;

(2) a calculation for the most probable distribution among the cells in this interval based on the notion of indistinguishable photons.

---

[3] "It [derivation of Planck's formula] was obtained from the condition that the quantum theoretic partition of states of the internal energy of the molecules is established only by the emission and absorption of radiation" (Einstein 1916, 47–62).

[4] The situation is even graver when considering the helium atom. Here, the energy required to raise the electron from the ground state atom to its first excited state is $\sim 20$ eV.

[5] This is set out in detail in Clarke (2017, §4.9).

The essence of (2) lies in the meaning Bose gives to 'distribution'. This turns out to be a count of the numbers of cells containing $0, 1, 2, \ldots$ photons assigned to the variables $p_0, p_1, p_2 \ldots$. The number of available cells $A$ is determined in (1) by the given frequency interval. There are also constraints: the total number $N$ of available quanta, and the fixed total energy $E$. So, the problem is to ascertain the number of ways that each set of values $p_0, p_1, p_2 \ldots$ can be assigned subject to these constraints.

Let us first consider the Bose counting procedure for the case $N = 2$ and $A = 2$, where the two combinations or distributions for sharing out the $N$ quanta into the $A$ cells are $(0, 2)$ and $(1, 1)$.[6] Table 4.1 sets out the $p_r$, for example, the first row containing values shows that there is one cell with zero photons, zero cells with one photon, and one cell with two photons.

The number of possible arrangements $n$ of the $N$ photons among the $A$ cells for each distribution is calculated from

$$n = \frac{A!}{p_0! p_1! p_2! p_3! p_4! \ldots} \tag{4.5}$$

where $p_0$ is the number of cells with zero photons, $p_1$ is the number with one photon, and so on. For example, using Table 4.1, the combination $(0, 2)$ of the number of photons in the two cells can occur in $2!/(1!0!1!) = 2$ ways as seen from arrangements 1 and 2 in Table 4.2, where a photon is indicated by a black spot. However, the distribution $(1, 1)$ has only one way of occurring which to Bose suggested that the photons must be regarded as indistinguishable. The distribution with

Table 4.1    The number of cells $p_r$ containing $r$ quanta for $N = 2$ quanta and $A = 2$ available cells.

| Distribution | $p_0$ | $p_1$ | $p_2$ |
|---|---|---|---|
| $(0, 2)$ | 1 | 0 | 1 |
| $(1, 1)$ | 0 | 2 | 0 |

[6]So, $(0, 2)$ means one cell has quanta and the other has 2.

Table 4.2  The Bose method of distributing $N = 2$ photons among $A = 2$ cells based on the assumption of indistinguishable photons shown as black spots: lines 1 and 2 show the combination of number of photons in cells $(0, 2)$ and line 3 gives $(1, 1)$.

|   | Cell 1 | Cell 2 |
|---|--------|--------|
| 1 | • •    | —      |
| 2 | —      | • •    |
| 3 | •      | •      |

Table 4.3  The Boltzmann method of distributing $N = 2$ photons among $A = 2$ cells based on the assumption of distinguishable photons.

|    | Cell 1 | Cell 2 |
|----|--------|--------|
| 1  | $P_1 P_2$ | —   |
| 2  | —      | $P_1 P_2$ |
| 3a | $P_1$  | $P_2$  |
| 3b | $P_2$  | $P_1$  |

the maximum number of possible arrangements subject to the total number and energy constraints is the most probable distribution for the given frequency interval.

If we consider the Boltzmann counting method for the same problem, the identity of each photon is retained. Let the two photons be denoted by $P_1, P_2$. The possible arrangements are given in Table 4.3 where we note that arrangement 3 in Table 4.2 can now occur in two different ways, 3a and 3b.

Both Bose and Einstein assume that the momentum of a light quantum is a vector with magnitude $p = h\nu/c$, but Bose departs from his predecessor by introducing a 6-D phase space in which the photon state is described at any instant by a point $(p_x, p_y, p_z, x, y, z)$. This is the calculation (4.1). A cell occupies a volume $h^3$ in this phase space so that, after taking account of the two polarization directions, the number of cells available in the $s$th frequency interval $(\nu_s, \nu_s + d\nu_s)$ is given by

$$A_s = \frac{2}{h^3} \int dp_x dp_y dp_z dx dy dz = \frac{8\pi \nu_s^2}{c^3} d\nu_s \cdot V \qquad (4.6)$$

Let us define $p_{qrs}$ as the number of cells containing $r$ quanta for the $q$th distribution in the $s$th frequency interval so that the total number of available cells is given by

$$A_s = \sum_r p_{qrs} \qquad (4.7)$$

and the total number of quanta to be distributed among these cells is

$$N_s = \sum_r r p_{qrs} \qquad (4.8)$$

The constraint on the constant total energy $E_s$ at frequency $\nu_s$ is

$$E_s = h\nu_s N_s = h\nu_s \sum_r r p_{qrs} \qquad (4.9)$$

Recalling (4.5), and using our modified notation, the number of ways a particular distribution $q$ can occur is given by

$$\frac{A_s!}{p_{q0s}! p_{q1s}! p_{q2s}! \dots} \qquad (4.10)$$

The total number of distributions is the sum of (4.10) over all $q$ and is a constant

$$C_s = \sum_q \frac{A_s!}{p_{q0s}! p_{q1s}! p_{q2s}! \dots} \qquad (4.11)$$

The probability $W_{qs}$ of a distribution $q$ at a particular frequency $\nu_s$ then becomes

$$W_{qs} = \frac{1}{C_s} \cdot \frac{A_s!}{p_{q0s}! p_{q1s}! p_{q2s}! \dots} \qquad (4.12)$$

The entropy $S_{qs}$ of the $q$th distribution for the quanta at frequency $\nu_s$ is to follow Boltzmann's law

$$S_{qs} = k \ln(W_{qs}) \tag{4.13}$$

We now aim to maximize the entropy (4.13) with respect to the $p_{qrs}$ by varying $q$ subject to the constraints (4.8) and (4.9). Applying the variation to (4.13) gives

$$\frac{\delta(S_{qs})}{\delta p_{qrs}} = k\frac{\delta(\ln(W_{qs}))}{\delta p_{qrs}} = -k\sum_r (\ln p_{qrs} + 1) = 0 \tag{4.14}$$

having used Stirling's approximation $p_{qrs}! \approx p_{qrs}^{p_{qrs}}$ with $A_s$ and $C_s$ as constants. Varying (4.7) and (4.9) accordingly produces

$$\frac{\delta(A_s)}{\delta p_{qrs}} = \sum_r 1 = 0 \tag{4.15}$$

and

$$\frac{\delta(E_s)}{\delta p_{qrs}} = h\nu_s \sum_r r = 0 \tag{4.16}$$

noting that the zeroes arise from the constancy of $E_s$ and $A_s$ as the distribution is varied. The use of undetermined multipliers $\gamma_1$ and $\gamma_2$ permits (4.14), (4.15), and (4.16) to be combined as

$$\sum_r (\ln p_{qrs} + 1 + \gamma_1 + \gamma_2 r h\nu_s) = 0 \tag{4.17}$$

to yield the most probable distribution (denoted by $q$) as

$$p_{qrs} = B_s e^{-\gamma_2 r h\nu_s} \tag{4.18}$$

where $B_s$ is constant. Using (4.7),

$$A_s = B_s \sum_{r=0} e^{-\gamma_2 r h\nu_s} = B_s(1 - e^{-\gamma_2 h\nu_s})^{-1} \tag{4.19}$$

Returning to (4.8), we also have

$$N_s = \sum_r r p_{qrs} = B_s \sum_{r=0} r e^{-\gamma_2 r h\nu_s} = B_s e^{-\gamma_2 h\nu_s}(1 - e^{-\gamma_2 h\nu_s})^{-2}$$

$$= A_s e^{-\gamma_2 h\nu_s}(1 - e^{-\gamma_2 h\nu_s})^{-1} = A_s(e^{\gamma_2 h\nu_s} - 1)^{-1} \tag{4.20}$$

Finally, (4.6) and (4.20) for a photon of energy $h\nu$ give

$$E_s = \frac{8\pi h\nu_s^3}{c^3} d\nu_s \cdot \frac{1}{e^{\gamma_2 h\nu_s} - 1} \cdot V \tag{4.21}$$

where $\gamma_2 = 1/kT$.

Bose had managed to avoid the problem that had plagued previous derivations, namely, the use of Maxwell's theory of electrodynamics, and replaced it with a 6-D phase space having a basic element of size $h^3$.

The Planck law will now be derived on the basis of the HSD theory. For this, there will be three steps:

(1) for the frequency interval $(\nu, \nu + d\nu)$, an expression for the number of cells in this interval;
(2) for the frequency interval $(\nu, \nu + d\nu)$, a calculation for the component of directed momentum striking the target surface;
(3) a calculation for the most probable distribution among the cells in this interval using the Bose counting method, but based on distinguishable photons.

## 4.3   Cells Represented as Direction Vectors

### 4.3.1   *The number of cells in $(\nu, \nu + d\nu)$*

With the notion that a photon is a single-wavelength transmission (spin-1), it is possible to use the HSD model to derive the number of cells that occupy volume $V$ in the frequency interval $(\nu, \nu + d\nu)$. Let us return to Figure 2.6 which shows a circularly polarized ray in a representational guide tube inducing a rotation with frequency $\nu$ in a perpendicular stationary plane it passes through. Let the velocity vector component $v_\perp$ be perpendicular to its propagation direction, and $v_\parallel$ parallel to it. For circularly polarized light with rake $\pi/4$, we have $v_\perp = v_\parallel = c$ so that the wavelength $\lambda$ is related to the radius of rotation $r_1$ by $\lambda = 2\pi r_1$; see (2.10) and (2.11). The area of the end of this vortex tube of radiation is

$$\pi r_1^2 = \pi \left(\frac{\lambda}{2\pi}\right)^2 = \frac{c^2}{4\pi\nu^2} \tag{4.22}$$

so that the number of wavelengths passing through unit area of the target surface is $4\pi\nu^2/c^2$. If we divide this by $\lambda$, the length of the vortex tube, the number of wavelengths per unit volume $N$ is given by

$$N = \frac{4\pi\nu^3}{c^3} \tag{4.23}$$

The number of cells that occupy volume $V$ in the frequency interval $(\nu, \nu + d\nu)$ is then

$$dN = \frac{24\pi\nu^2}{c^3}d\nu \tag{4.24}$$

having taken into consideration the two circular polarization senses.

### 4.3.2 *Tubes incident at an angle $\theta$*

It is clear that a tube has directed momentum $h\nu/c$ which means that consideration must also be given to the orientation. So, the next task is to consider how this affects rays incident on the target area.

In Figure 4.1, the target area is ABCD and the component of momentum parallel to the surface normal $\hat{n}$ is $h\nu\cos\theta/c$, so this momentum reduction must be considered. Also, in the calculation of (4.22), it was assumed that the vortex tube ran parallel to $\hat{n}$ and that the tube cross-sectional area that fell on the target area ABCD was circular. However, for $0 < \theta < \pi/2$, it should be elliptical with area $\pi ab$, where $a$ is its half-major axis parallel to AB, and $b$ is its half-minor axis parallel to BC. In that case, $b = r_1$ is unaffected but $a = r_1/\cos\theta$, so in (4.22) we should have $\pi r_1^2 \to \pi r_1^2/\cos\theta$.

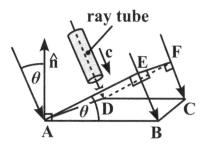

Figure 4.1    A ray tube incident on area ABCD at an angle $\theta$.

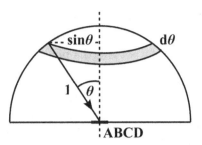

Figure 4.2    Hemisphere of rays centered on ABCD at O.

Including both the momentum and cross-sectional area dependency on $\theta$, (4.24) becomes as follows:

$$\mathrm{d}N = \frac{24\pi\nu^2}{c^3}\mathrm{d}\nu\cos^2\theta \qquad (4.25)$$

The proportion of the total radiation reaching ABCD depends on the angle of incidence $\theta$. In Figure 4.2, a hemisphere of unit radius has its center located on the element ABCD. The area of the strip is $2\pi\sin\theta\,d\theta$ and the surface area of the unit-radius hemisphere is $2\pi$.

This means that the fraction of the total number of rays $dN$ in the interval $(\nu, \nu + \mathrm{d}\nu)$ passing through this strip toward area ABCD from the interval $(\theta, \theta + d\theta)$ is

$$\frac{2\pi\sin\theta\,d\theta}{2\pi} = \sin\theta\,d\theta \qquad (4.26)$$

The total number of entrances to ABCD available for the rays in volume $V$ now results from summing all strips over $\theta$, so that using (4.25) and (4.26), we arrive at the total number of cells

$$A_s = dN = \frac{24\pi\nu^2}{c^3}\mathrm{d}\nu\,V\int_0^{\frac{\pi}{2}}\cos^2\theta\sin\theta\,d\theta = \frac{8\pi\nu^2}{c^3}\mathrm{d}\nu\,V \qquad (4.27)$$

To obtain Planck's law, we now substitute (4.27) into (4.20) and multiply by $h\nu$, noting that its $\cos\theta$ dependency in relation to its momentum $h\nu/c$ has already been accounted for in (4.25).

The greater the $\theta$ for a cell, the smaller the contribution its momentum makes to the energy density. This lends us the insight that for each interval $(\nu, \nu + \mathrm{d}\nu)$ in which all photons have approximately identical momentum, each cell corresponds to a unique orientation or vector. This notion becomes apparent in the Bose phase space approach when we pass from Cartesian to spherical polar coordinates. The following section will consider how the Bose counting method can arise from the assumption of *distinguishable* photons.

## 4.4 Bose Counting with Distinguishable Photons

### 4.4.1 *The continuous cavity path*

Hitherto, the notion that photons are observationally *indistinguishable* has been the only assumption from which the Bose counting method has arisen, and it has been cited in support of the Heisenberg position that

> it seems more reasonable to try to establish a theoretical quantum mechanics, analogous to classical mechanics, but in which only relations between observable quantities occur.[7] (Heisenberg 1925, 880)

However, a novel procedure will now be introduced in which it will be shown how the Bose counting procedure can be carried out using *distinguishable* photons. The foundation of our analysis shall be rotating sequences in a closed cavity.

First, let us introduce a rotating sequence as one in which its elements are arranged in a circle and the order is viewed in one direction, say, clockwise. The first point to be established concerns the invariance of a rotating sequence after an exchange of different elements which can only occur when the number of elements in the sequence $N = 2$. To see this, consider the rotating sequence $P_2 P_1 P_2 P_1 \ldots$ (read from right to left) formed by placing one element $P_1$ at the top of a circle and the other $P_2$ at the bottom and moving clockwise. This turns out to be equivalent to $P_1 P_2 P_1 P_2 \ldots$ after juxtaposing $P_1$ and $P_2$, because $P_2$ still follows $P_1$ even after the exchange. However, with $N = 3$ elements arranged to produce the clockwise sequence $P_1 P_3 P_2 P_1 P_3 P_2 P_1 \ldots$, this does

---

[7]  *"und zu versuchen, eine der klassischen Mechanik analoge quantentheoretische Mechanik auszubilden, in welcher nur Beziehungen zwischen beobachtbaren Grössen vorkommen."*

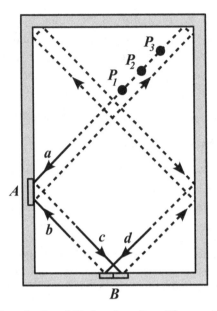

Figure 4.3    Perfectly reflecting 2-D closed cavity with a continuous photon path. Each reflector $A$, $B$ has two cells, defined by the incident direction, for the $N = 3$ photons to occupy. The cells follow a rotating sequence $c, d, b, a, c, \ldots$ (read from left to right). The passing photons $P_3, P_2, P_1$, (read from right to left) are shown as black circles and also follow a rotating sequence viewed from a stationary point on the path.

not remain the same after interchanging $P_2$ and $P_3$. It instead becomes $P_1 P_2 P_3 P_1 P_2 P_3 P_1 \ldots$. Since $P_2$ no longer follows $P_1$, then the invariance is lost. This holds for $N \geq 3$ when any two different elements of a rotating sequence are interchanged.

Let us now consider a closed rectangular 2-D perfectly reflecting cavity in which $N = 3$ photons in the frequency interval $(\nu, \nu + \Delta\nu)$ are passing along a continuous closed path; see Figure 4.3. The photons strike reflector $A$ at $a$ and $b$, and an adjustable reflector–detector $B$ at $c$ and $d$. Each direction of approach (vector) to a surface of interest in this frequency interval is to be identified with a unique cell, and in our present example each surface of interest has two cells that a photon might inhabit. Due to the continuous path, the order of visitation of the cells in the cavity is a rotating sequence $c, d, b, a, c, \ldots$ (read from left to right).

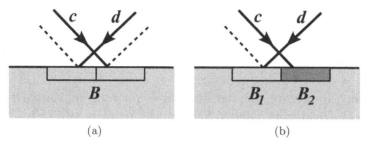

Figure 4.4    Detail of an adjustable surface B showing how it can be switched from (a) a perfectly reflecting flat surface B into (b) a reflector $B_1$ and a detector $B_2$.

Let us also introduce a detail into surface B that at any moment it can be switched from a perfectly reflecting surface into $B_1$ a reflector, and $B_2$ a detector; see Figure 4.4. The former can only receive and reflect a photon approaching in cell $d$ while the latter can register and identify one approaching in $c$. Our intention is to allow the photon sequence to perform several circuits before switching to $B_2$. Allow also that, using SPDC, $\Delta\nu$ can be chosen for the frequency interval $(\nu, \nu + \Delta\nu)$ in such a way that the three photons $P_1, P_2, P_3$ have distinguishable frequencies $\nu_1, \nu_2, \nu_3$ in that interval, ones that detector $B_2$ can distinguish. This means that we can not only know that cell $c$ is occupied but we can also identify which of the three photons is occupying it. However, the identity of this photon is not known in advance of the detector's registration. We are now ready to illustrate our new method.

The adjustable surface $B$ is initially set as a double reflector; see Figure 4.4(a). In rapid succession, we pump the three photons $P_3, P_2, P_1$ (read right to left) into the perfectly reflecting cavity. If we are to obtain the desired sequence, the task must be completed before $P_1$ returns to its entry point. After some arbitrary time interval has elapsed, $B$ is switched to its dual mode as shown in Figure 4.4(b). The first photon that registers at $B_2$ is identified and the possible distributions of the remaining two photons between the remaining three cells are calculated from the fact that both the cell and photon sequences are rotational. This calculation will now be examined in detail.

### 4.4.2   *Examples of distributions*

4.4.2.1   *Example 1: Three photons, four cells*

(a)  Possible distributions

Referring to Figure 4.3, let the three photons in the cavity be $P_3, P_2, P_1$ (read from right to left) and let us place two cells $a$ and $b$ at A, and the two cells $c$ and $d$ at B. Apart from the surface at cell $c$ which can be switched from a reflector to a detector, the other three cells impinge on reflectors. Consider $P_3$ to be detected and identified at $c$. In Table 4.4, we place $P_3$ in each row in the column under $c$. Our task is to distribute the remaining photons $P_1$ and $P_2$ among the cells a, b, and d while preserving their rotation sequences. The cell sequence is guaranteed by the table layout (read right to left) and is the order the photons visit them. For the number of photons in each remaining cell, the possible combinations are $(0, 0, 2)$ or $(0, 1, 1)$, and the possible distributions are shown in Table 4.4.

(b)  Bose result

The expected Bose result using (4.11) for the sum of the distributions with $A_s = 3$ as the number of cells, and photon number combinations

Table 4.4   With $P_3$ assigned to cell $c$, there are six ways of distributing $P_1$ and $P_2$ among the remaining cells $a, d, b$ while retaining the rotation sequence of the photons.

|   | Detector A cell a | Detector B cell c | Detector B cell d | Detector A cell b |
|---|---|---|---|---|
| 1 | $P_2 P_1$ | $P_3$ | — | — |
| 2 | — | $P_3$ | $P_2 P_1$ | — |
| 3 | — | $P_3$ | — | $P_2 P_1$ |
| 4 | — | $P_3$ | $P_2$ | $P_1$ |
| 5 | $P_1$ | $P_3$ | — | $P_2$ |
| 6 | $P_1$ | $P_3$ | $P_2$ | — |

$(0, 0, 2)$, $(0, 1, 1)$ is

$$\sum_q \frac{A_s!}{p_{q0s}! p_{q1s}! p_{q2s}!} = \frac{3!}{2!0!1!} + \frac{3!}{1!2!0!} = 6 \qquad (4.28)$$

So, if the photon sequence is known, then the identification of only one photon in a single cell is sufficient to effect a synchronization of the photon and cell sequences and identify the photons in each cell for each distribution. The concept of 'indistinguishable' is not necessary.

## (c) New counting method

We shall now obtain a formula that predicts the correct number of arrangements by examining the following algorithm applied to the present example. Consider $N = 3$ photons and $A = 4$ cells, and let us represent them by $N = 3$ spots and $A - 1 = 3$ partitions. The spaces between the partitions are the cells labeled in their order of visitation; see Figure 4.5. In Figure 4.5(a), we see the elements to be arranged: three photons and three partitions. Photon $P_3$ has been identified in cell $c$ and this is shown in Figure 4.5(b). In Figure 4.5(c), the partitions on each side of cell $c$ are collapsed to a single partition which is labeled $P_3$. The problem is now to arrange the remaining two spots and two partitions without regard to order. The six possible arrangements are shown in Figure 4.6, where we have reduced the problem to the arrangement of $(N - 1)$ photons and $(A - 2)$ partitions. The calculation for the sum is as follows:

$$\frac{(N + A - 3)!}{(N - 1)!(A - 2)!} = \frac{(3 + 4 - 3)!}{(3 - 1)!(4 - 2)!} = \frac{4!}{2!2!} = 6 \qquad (4.29)$$

There are six ways that two photons and two partitions can be arranged without regard to order and these are set out in Figure 4.6. In each case, the collapse of cell $c$ is now reversed, and Figure 4.7 shows the result when this is applied to the example Figure 4.6(e).

Finally, using $P_3$ as a datum, the photons are labeled using the rotating sequence $P_3, P_2, P_1, P_3, P_2, \ldots$ so that the result for

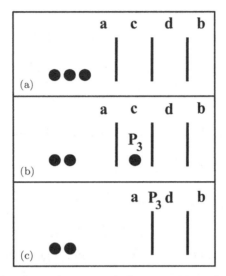

Figure 4.5  (a) The three photons and three partitions for arrangement. (b) Photon $P_3$ is assigned to cell $c$. (c) The partitions on each side of $P_3$ are collapsed to a single partition which is labeled $P_3$, leaving two photons and two partitions for arrangement.

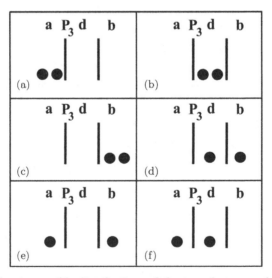

Figure 4.6  The six possible distributions of the two photons and two partitions without regard to order.

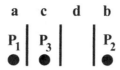

Figure 4.7   The result of reversing the collapse of cell *c* in Figure 4.6(e).

Figure 4.6(e) is shown in Figure 4.7 as $P_1, P_3, -, P_2$. This is also row 5 in Table 4.4.

### 4.4.2.2  *Example 2: Five photons, five cells*

#### (a)  Possible distributions

We now add a level of complexity. Figure 4.8 shows a five-cell cavity design where the mirrors $M_1$ and $M_2$ completely reflect approaching photons, and the surfaces $A$ and $B$ are initially completely reflecting. Surface $B$ can be switched at any moment from a completely reflecting flat surface into $B_1$ a reflector at cell *d*, and $B_2$ a detector at cell *b*. The cells are visited in the sequence is $a, b, c, d, e, a, \ldots$ (read from left to right) by the photon sequence $P_1, P_5, P_4, P_3, P_2, P_1, \ldots$ (read from right to left) which is the order that the photons approach the cells.

Allow $P_4$ to be detected at cell *b*. What remains is to distribute the remaining four photons among the four cells *a, c, d, e* while maintaining their rotation sequences. The possible combinations of distributions for these cells are (0, 0, 0, 4), (0, 0, 1, 3), (0, 0, 2, 2), (0, 1, 1, 2), and (1, 1, 1, 1). As shown in Example 1, once the possible distributions have been set out, the photons can be labeled (distinguished) by reference to the order of their rotating sequence.

#### (b)  Bose result

Using (4.11), the expected Bose result for the number of these distributions is

$$\frac{4!}{3!0!0!0!1!} + \frac{4!}{2!1!0!1!0!} + \frac{4!}{2!0!2!0!0!} + \frac{4!}{1!2!1!0!0!} + \frac{4!}{0!4!0!0!0!} = 35$$

$$(4.30)$$

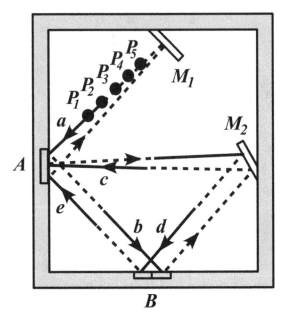

Figure 4.8   Perfectly reflecting closed cavity with a continuous photon path maintained by mirrors $M_1$ and $M_2$. Detector $A$ has three cells $a$, $c$, $e$, and detector $B$ has two cells $b$ and $d$, where $b$ can be switched to a detector. The cells follow a rotating sequence, $a, b, c, d, e, a \ldots$ (read from left to right). The photons $P_5, P_4, P_3, P_2, P_1$ (read from right to left) are shown as black spots, and also follow a rotating sequence viewed from a stationary point on the path.

## (c) New counting method

With the number of photons $N = 5$ and the number of cells $A = 5$, we get

$$\frac{(N + A - 3)!}{(N - 1)!\,(A - 2)!} = \frac{(5 + 5 - 3)!}{(5 - 1)!\,(5 - 2)!} = \frac{7!}{4!3!} = 35 \qquad (4.31)$$

## 4.5   Concluding Remarks

When we use photons that possess a known rotating sequence, on a closed path consisting of a rotating sequence of cell visitation, then since each photon is capable of being identified using a detector in one known cell, the photon and cell sequences can be synchronized. The set of possible distributions that arises follows the Bose–Einstein statistics and we have shown here that it need not depend on the notion of 'indistinguishable'. On the contrary, the photons in each cell are distinguishable and only a single photon needs to be identified in a known cell for the method to work.

Of course, so long as Bose confined his attention only to a single arbitrarily small absorption area of the cavity wall rather than the totality of cavity areas, he was prohibited from considering continuous photon-flow paths in the cavity, and from characterizing the cells as vectors, each with a unique direction. Without this wider view, it is understandable that a theory of rotating photon and cell sequences did not occur to him.

It is worth remarking that the HSD theory accords with the idea of multiple photons in a single cell. Since in the frequency interval $(\nu, \nu + d\nu)$ each cell has a unique orientation, then multiple directed tubes of the same frequency in a particular cell can be brought together as a longitudinal iteration, phase-matched end to end. A photon tube can then be $n\lambda$ in extent, where $\lambda$ is the length of one tube and $n \in Z^+$.

## References

Bose, N. 'Planck's Gesetz und Lichtquantenhypotheses'. *Zeitschrift für Physik*, 26 (1924): 178–181.

Clarke, B. R. *The Quantum Puzzle: Critique of Quantum Theory and Electrodynamics*. World Scientific Publishing, 2017.

Einstein, A. 'Über einen die Erzeugung und Verwandlung des Lictes betreffenden heuristischen Gesichtspunkt'. *Annalen der Physik*, 17 (1905): 132–148. English translation 'On a heuristic point of view about the creation and absorption of light' in A. B. Arons, and M. B. Peppard, *American Journal of Physics*, 33 (1965): 367–374.

Einstein, A. 'Zur Quantentheorie der Strahlung'. *Mitteilungen der Physikalischen Gessellschaft, Zürich*, 18 (1916): 47–62.

Heisenberg, W. 'Über quantentheoretische Umdeutung kinematischer und mechanischer Beziehungen'. *Zeitschrift für Physik*, 33 (1925): 879–893; English translation, 'Quantum-theoretical re-interpretation of kinematical and mechanical relations,' in B. L. van der Waerden, *Sources of Quantum Mechanics*. Amsterdam: North-Holland Publishing Company, 1967.

Jeans, J. 'A comparison between two theories of radiation'. *Nature*, 72 (1905): 293–294.

Kuhn, T. *Black-body Theory and the Quantum Discontinuity, 1894–1912*. Oxford University Press, 1978.

Mott, N. 'On teaching quantum phenomena'. *Contemporary Physics*, 5 (1964): 401–408.

Planck, M. 'Über irreversible Strahlungvorgänge'. *Annalen der Physik*, 1 (1900): 69–122.

Rubens, H., and F. Kurlbaum. 'Anwendung der Methode der Reststrahlen zur Prufung des Strahlungsgesetzes'. *Annalen der Physik*, 4 (1901): 649–666.

# Chapter 5

# The Helical Array Dislocation

*A single-photon front, consisting initially of an array of helical space dislocation (HSD) of identical phase, can be given optical orbital angular momentum (OAM) by passing it through an optical element and rendering the phase dependent on the azimuthal angle. The rotation of the Poynting vector about the optic axis results in a helical front.*

## 5.1 Overview

So far, we have examined the ideas behind the circularly polarized ray and have characterized it as a traveling non-rotating screw thread or helical space dislocation (HSD). The optical spin angular momentum (SAM) rotation it induces in a stationary plane set perpendicular to its path has been designated 'spin-1'.[1] The Grangier experiment and similar ones have also been explored to argue that spin-1 is iterated transversely as an array of traveling HSD. In contrast, the phenomenon of heat radiation is assumed to rely on a longitudinal iteration of the HSD. The task now is to vary the phases azimuthally of the HSD on the array and produce optical orbital angular momentum (OAM) in the form of a helical front.

In the last 25 years, several experiments have been conducted with laser-sourced Gaussian $TEM_{00}$ beams modified with an optical element to produce OAM. A small-angle divergence of the laser beam in relation to the optical axis allows what is known as a paraxial-limit solution of the wave equation in the form of rectangularly symmetric Hermite-polynomial modes with a Gaussian

---

[1]This is a new nomenclature.

amplitude profile (Hermite–Gaussian, HG, modes). A combination of such modes, taken from the components of a $HG$ mode rotated at 45° to the $x - y$ axes in the beam plane, have then been given $\pi/2$ relative phase shifts using various optical elements such as spiral wave-plates (Beijersbergen *et al.* 1994), holograms (Friese *et al.* 1996; He *et al.* 1995; Soskin *et al.* 1997; Tang *et al.* 1995),[2] and cylindrical lenses (Allen *et al.* 1992; Courtial and Padgett 1999; Padgett *et al.* 1996; Simpson *et al.* 1997). This produces a Laguerre–Gaussian (LG) beam with a helical phase structure (OAM) that retains a constant beam waist.[3] The azimuthal phase dependence has the form $\exp(il\phi)$ resulting in an OAM of magnitude $l\hbar$, and if we assign the propagation axis along $z$, the field amplitude takes the form $\psi(r, \phi, z) = \psi_o(r, z) \exp(il\phi)$. In other words,

> Light beams with OAM have a 'twisted' or helical [constant-] phase structure, where the phase [at constant $z$, varies as one] winds azimuthally around the optic axis. (Krenn *et al.* 2017)

Destructive interference ensures that the beam center has a dark spot with zero intensity.

Although the intensity of the beam is taken to be a continuous function of the spatial coordinates, we investigate here the idea that a HSD model of a photon front might still be admissible by deconstructing the beam into an array of HSD tubes ranged across the beam front with phase variation. The behavior of OAM beams consisting of an aggregate of photons is reviewed under various conditions to lend insight as to the behavior of the *individual* photon under similar conditions.

## 5.2    Beam Parameters

### 5.2.1    *Characteristics of HG and LG modes*

We now set out the theory of HG and LG beams as given in the modern literature. A selection of $HG_{mn}$ modes is shown in Figure 5.1.

---

[2]The hologram has a helical phase structure.
[3]The mathematical solutions have been presented by Abramochkin and Volostnikov (1991).

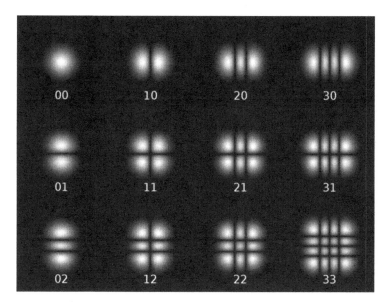

Figure 5.1   Twelve Hermite–Gaussian $HG_{mn}$ modes.

Index $m$ gives the number of dark columns and index $n$ denotes the number of dark rows. In Figure 5.2, a number of $LG_{lp}$ modes are shown. Here, $l$ is the OAM as a multiple of $\hbar$ and is a measure of the central void diameter, while also indicating the number of $2\pi$ phase cycles taken around the beam at a constant radius. The index $p$ is the number of dark rings, and $p + 1$ is the number of bright rings or nodes. We note that $l = n - m$ and $p = m$.

Figure 5.3 shows a diagonal $HG_{02}$ mode (left) decomposed into three components which are then recombined with a relative phase shift of $\pi/2$ (indicated by $i = \sqrt{-1}$) between adjacent modes to form $LG_{20}$ in Figure 5.4.[4]

The output from a laser with non-cylindrical symmetry is usually modeled by HG amplitude modes. The lowest mode is $HG_{00}$ which is a single Gaussian spot when brought into focus, and if higher $HG_{mn}$ modes are to be obtained from this, a cross wire can be positioned in

---

[4]For similar diagrams showing the construction of the $LG_{10}$ mode from the components of a diagonal $HG_{10}$, see Padgett *et al.* (1996, Fig. 2).

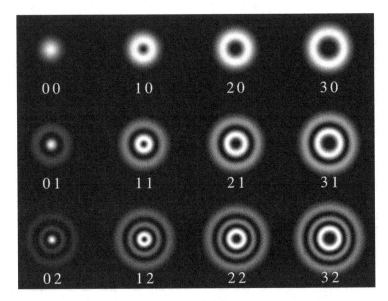

Figure 5.2    Twelve Laguerre–Gaussian $LG_{lp}$ modes.

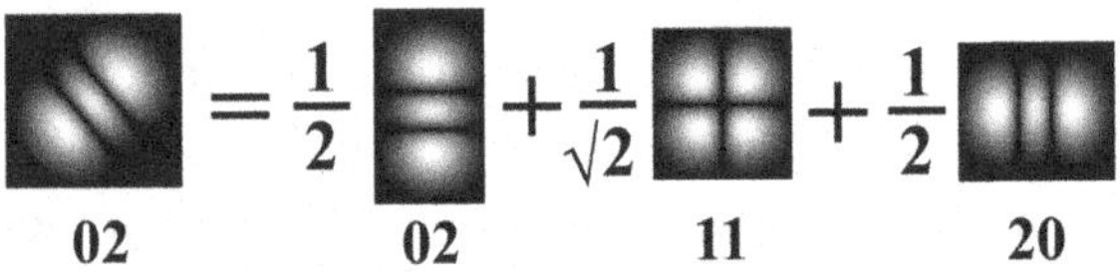

$$\boxed{}\ \ 02\ =\ \frac{1}{2}\ \boxed{}\ \ 02\ +\ \frac{1}{\sqrt{2}}\ \boxed{}\ \ 11\ +\ \frac{1}{2}\ \boxed{}\ \ 20$$

Figure 5.3    $HG_{02}$ diagonal mode construction from three in-phase HG modes.

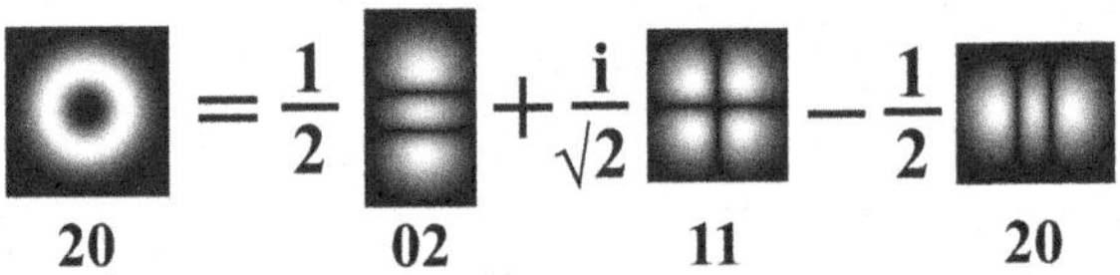

$$\boxed{}\ \ 20\ =\ \frac{1}{2}\ \boxed{}\ \ 02\ +\ \frac{i}{\sqrt{2}}\ \boxed{}\ \ 11\ -\ \frac{1}{2}\ \boxed{}\ \ 20$$

Figure 5.4    $LG_{20}$ (angular momentum $l = 2$) mode construction from three HG modes, with each term possessing a $\pi/2$ phase advance (multiplication by $i$) in relation to the previous mode.

the laser cavity so that the wires are in alignment with the desired nodes (dark lines in Figure 5.1).[5] This has the effect of breaking the

---

[5]For example, Padgett *et al.* (1996, 80) have used a tungsten cross wire with diameter $10\,\mu m$.

output intensity of the laser into separate spatial components. When the beam is focused by a lens, it has a minimum waist parameter $w_0$ at the focus $z = 0$, which is the radius from the beam axis at which the amplitude falls to $e^{-1}$ of its maximum value. It also has a Rayleigh range $z_R$ — the distance from the focus along the $z$ axis at which the waist has increased to $\sqrt{2}w_0$ — which is given by $z_R = \pi w_0^2/\lambda$, where $\lambda$ is the beam wavelength. The waist $w$ at a distance $z$ from the focus is given by $w(z) = w_0\sqrt{1 + (z/z_R)^2}$.

### 5.2.2 *Gouy phase shift*

Consider a plane wave impinging on a convex lens. The rays in the wave are redirected, and any converging light wave that passes through a focus experiences a Gouy phase shift compared to that of a plane wave of the same frequency as it propagates from far field to far field. Here, the rays traveling in different directions acquire different phase shifts when viewed in a plane parallel to the original beam.[6] For a $HG_{mn}$ mode, where $m$ and $n$ are indices for the $x$ and $y$ directions, and with identical Rayleigh ranges in the $x - z$ and $y - z$ planes, the Gouy phase is

$$\psi(z) = (m + n + 1)\tan^{-1}(z/z_R) \tag{5.1}$$

while for a $LG_{lp}$ mode,

$$\psi(z) = (2p + l + 1)\tan^{-1}(z/z_R) \tag{5.2}$$

However, for a cylindrical lens, the Rayleigh ranges in the $x - z$ and $y - z$ planes are different, so that

$$\psi(z) = \left(m + \frac{1}{2}\right)\tan^{-1}(z/z_{R_{x-z}}) + \left(n + \frac{1}{2}\right)\tan^{-1}(z/z_{R_{y-z}}) \tag{5.3}$$

---

[6]The Gouy phase shift was reported as long ago as 1890, and constitutes a total $n'\pi/2$ axial phase shift, where $n' = 1$ for a line focus in a cylindrical wave and $n' = 2$ for a point focus in a spherical one; see Feng and Winful (2001). The phase shifts can only be due to path differences and a retardation that is determined by the lens thickness traversed.

This allows for a difference in focusing along the $x$ and $y$ axes resulting in an elliptical Gaussian beam. For example, if the cylindrical lens axis is aligned with the $x$ axis, then only the Rayleigh range $z_{R_{y-z}}$ is affected, so that when the modes $HG_{mn}$ and $HG_{nm}$ ($m \neq n$) are passed through the lens, their Gouy phase shifts will differ. This property can be exploited to generate a $LG_{10}$ mode in which $HG_{10}$ and $HG_{01}$ must be $\pi/2$ out of phase.

### 5.2.3  *Cylindrical mode converter*

We now consider the action of a cylindrical mode converter. For a given cylindrical lens of focal length $f$, and a diverging beam input with wavefront radius $R_i$ and waist $w$ at its origin ($z = 0$), it is possible to calculate the focal length $f_{in}$ of the spherical lens required to prepare the beam for entrance into the converter. First, the separation of the two cylindrical lenses must be set at $2d = \sqrt{2}f$ while the incident beam must have Rayleigh range $z_r = (1 + 1/\sqrt{2})f$. This is ensured by selection of a spherical lens of focal length $f_{in}$ given by

$$\frac{1}{f_{in}} = \frac{\sqrt{2kw^2 z_r - 4z_r^2}}{kw^2 z_r} + \frac{1}{R_i} \tag{5.4}$$

Padgett *et al.* (1996, 80) have reported an experiment using $f = 25.4\,\text{mm}$, $2d = 35.9\,\text{mm}$, and $z_r = 43.3\,\text{mm}$. Trial and error has informed their choice as to the focal length for the input lens to the mode converter, so that $f_{in} = 250\,\text{mm}$. In Figure 5.5, the layout of a cylindrical lens mode converter is shown, where the components of a diagonal $HG_{10}$ undergo relative phase shifts to create a $LG_{10}$ mode.

In order to obtain a phase reference for the intensity pattern and obtain a meaningful beam profile on the viewing screen, beam splitters are used to redirect both the output from the mode converter and the original near-planar wavefront of the HG input mode, into a Mach–Zender interferometer. The consequent interference profiles are given in Figure 5.2.

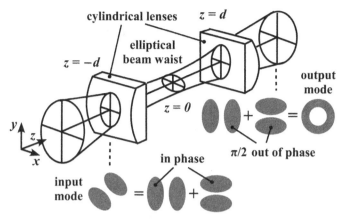

Figure 5.5    Production of a $LG_{10}$ mode by creating a relative phase shift of $\pi/2$ in the components of a diagonal ($\pi/4$ to $x$ axis) $HG_{10}$ mode using a cylindrical lens mode converter (Padgett *et al.* 1996, Fig. 4).

## 5.3    Beam Angular Momentum

It is possible to generate a LG beam of rays that carries $(l + \sigma)\hbar$ total angular momentum (OAM and SAM) per photon. Here, the $\sigma = 1$ results from left-circularly polarized, $\sigma = -1$ from right-circularly polarized rays, and $\sigma = 0$ from linearly polarized rays. He *et al.* (1995) have used a linearly polarized source to show that trapped absorptive particles will rotate in the same direction as the OAM beam that impinges upon them, and will change direction when the beam sense is reversed. They later observed that on changing the beam from linearly polarized to circularly polarized, there was an increase in the rotation frequency of the fragment when the spin and orbital rotations had the same sense, but a decrease when they had the opposite sense; see also Friese *et al.* (1996). A more precise observation that when the total angular momentum of the beam is doubled, the rotation frequency of the trapped absorbing particle also doubles has been reported by Simpson *et al.* (1997). This suggests an energy transfer.

There is also the question of angular momentum exchange when light passes *through* a medium. For the case of SAM, to transform right- $(-\hbar)$ into left-circularly polarized light $(+\hbar)$, a suspended birefringent plate takes on a clockwise torque (viewed in the ray direction), while for OAM, the transformation of a LG beam with $-l\hbar$ into one with $+l\hbar$ similarly creates a clockwise torque on passing through suspended cylindrical lenses (Allen *et al.* 1992, Fig. 1). This suggests angular momentum conservation.

Tang *et al.* (1995) have remarked that

All Laguerre–Gaussian modes [LG$_{lp}$] with $p = 0$ and $l \neq 0$ are helical waves, each with a charge $l$ phase singularity, and with equiphase surfaces screwing around the cavity axis with $\lambda/l$ pitch.

### 5.4   Single-Photon OAM

Allen *et al.* (1992, 8189) have suggested that a *single* photon is capable of carrying the total OAM; see Figure 5.6:

We have demonstrated [theoretically] that a Laguerre–Gaussian laser mode has a well-defined OAM equal to $l\hbar$ per photon, with $l$ the azimuthal mode index.

Consider a plane wave beam from a laser composed of an aggregate of circularly polarized photons having equal phase. When passed through a suitable mode converter, the OAM beam that

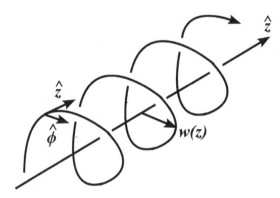

Figure 5.6   The spiraling curve represents the trajectory of the Poynting vector of a linearly polarized Laguerre–Gaussian mode of waist radius $w(z)$ from Allen *et al.* (1992, Fig. 2).

emerges contains these same photons in a well-collimated beam (negligible divergence), but now certain relative phase shifts have been induced as we rotate azimuthally around the optic axis.[7] In a given stationary plane perpendicular to the beam, it is the phase relationship between contiguous circularly polarized rays as we circulate around the beam axis at constant radius that creates the spiral amplitude and the OAM.

The representation of the trajectory of the Poynting vector given by Allen *et al.* (1992) is shown in Figure 5.6, but the Poynting vector only manifests itself when it interacts with the stationary plane that it passes through. The key question now is this. What does this OAM screw thread represent? Is it a non-rotating structure that only affects the plane when the whole thread advances? This would then mirror our model of the circularly polarized ray. Or, does it indicate the trajectory that a circularly polarized ray takes through space, one that is perpendicular to the screw-thread blade? In the former case, the rotation imparted to the plane will be opposite to the way the thread is wound. For the latter, the momentum of the circularly polarized ray imparts a rotation to the plane in the azimuthal direction that the ray strikes it; see Figure 5.7. We choose the second case here because it would be difficult to visualize how the Poynting vector could take on a helical trajectory if the rays actually producing the momentum do not.[8] In other words, the circularly polarized rays actually rotate around the optic axis.

Several methods have relied on attenuated beams to measure the OAM of 'single' photons (Wei *et al.* 2003). Leach *et al.* (2004) have obtained photographs of single-photon OAM states. Their experiment sorts OAM states (with $l = 0, 1, 2, 3, 4$) using an

---

[7]As Giovannini *et al.* (2015) have pointed out, a phase delay can be measured between time-correlated photons pairs simply by using one as a reference and extending the path traveled by the other using an axicon. An axicon is a conical lens that generates a ring-shaped beam profile.

[8]This following is stated by Simon (2016, 4): "The Poynting vector $\vec{S} = \vec{E} \times \vec{H}$ must be perpendicular to the wavefront, so it is at a non-zero angle to the propagation direction. $\vec{S}$ therefore, rotates about the axis as the wave propagates, leading to the existence of non-zero orbital angular momentum."

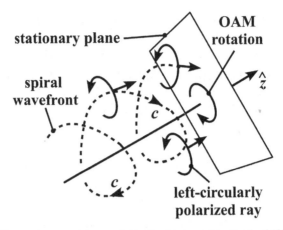

Figure 5.7   The trajectory (shown dotted) of tubes of individual HSD generating left-circular polarization (curved arrows). As the circularly polarized threads strike the plane, they generate a clockwise OAM rotation.

arrangement of Dove prisms (for the relative phase rotation of two beams),[9] beam splitters (to sort different states), and holograms (to shift even $l$ to odd values). Having previously used a helium–neon laser source at $<1\,\mathrm{mW}$ attenuated with neutral density filters to $<0.3\,\mathrm{nW}$ (Leach *et al.* 2002), they realized that their claim for using "intensities so low that on average less than one photon was present in each interferometer at any one time" had to be tempered with the confession that they "did not detect photons individually".

A more reliable method of obtaining a single-photon beam is spontaneous parametric down conversion (SPDC). In SPDC, a pumped photon beam is passed through a nonlinear crystal, such as beta-barium borate (BBO), and although most photons emerge from the crystal intact, occasionally a single photon is divided or down-converted into a photon pair, the members of which are referred to as the 'signal' and 'idler'.[10] If one of the pair is detected, the other member is also understood to have been created. Both energy

---

[9]A Dove prism introduces a phase shift that depends on the OAM of the rays and the rotation angle of the prism.

[10]"Approximately 1 in every $10^{12}$ photons are down-converted" (McLaren *et al.* 2015).

(sum of angular velocity magnitudes) and momentum (sum of vector wavenumbers) are conserved in the process, so the pair is locked in a conservation relationship. Various classifications of the process are possible. A Type I conversion means they share the same polarization, perpendicular to that of the pump, whereas a Type II conversion yields perpendicular polarizations, one of which has the same polarization as the pump. Also, in a degenerate down conversion, the two photons have the same wavelength, while a non-degenerate one produces different wavelengths. The orientation of the optic axis of the crystal relative to the laser pump allows the division into collinear and non-collinear processes, the latter producing a more distinct central dark spot for the conical distribution of photons emerging from the crystal and a greater spectrum of transverse modes. The radial intensity distribution increases as $\sqrt{l}$, and so as the pump waist increases, the number of possible OAM modes increases. Decreasing the crystal thickness also increases the number of modes.[11]

The first measurement of *single*-photon OAM using SPDC is commonly cited as that obtained by Mair *et al.* (2001). Their experiment sets out to show that the two photons detected in each pair production conserve OAM, and that in the process they exhibit entanglement. Their apparatus consists of an Argon-ion laser operating at 351 nm, capable of generating a simple Gaussian mode profile ($l = 0$) or a first-order $LG$ mode ($l = \pm1$) after passing the beam through a mode converter.[12] A BBO crystal with 1.5 mm thickness, cut for Type I phase matching, produces down-converted photons at 702 nm when the optic axis is at $4°$ to the axis of the laser. A laser operating at 632 nm was brought in to test the transmission efficiency of the computer-generated holograms.[13]

---

[11] See the section on 'bandwidth' in McLaren *et al.* (2015, Figure 4).

[12] Although not specified by the authors, this appears to have been a cylindrical lens arrangement.

[13] "The hologram is a phase grating with $\Delta m$ dislocations in the center [... and] the $n$th diffraction order [has] an index $l = n\Delta m$", Mair *et al.* (2001, Figure 2 caption). Of course, the aim is to reduce the LG mode to a Gaussian one.

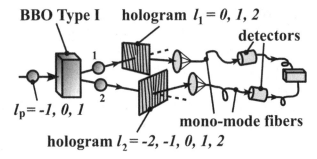

Figure 5.8    Experimental setup for single-photon mode detection. The hologram modifies the incoming LG mode into a Gaussian mode with a narrow focus that can be transmitted along a mono-mode fiber.

The first part of the experiment, the part that is of interest here, aims to show that the two SPDC photons that excite the coincidence detectors, having taken different paths, conserve the total OAM of the pump beam; see Figure 5.8.[14] The pump beam (modified by a mode converter where appropriate) has an OAM value of $l_p = -1, 0, 1$. After passing through the crystal, two photons are produced, one of which takes Path 1 and the other Path 2. The question arises as to what OAM each possesses. Path 1 is checked for modes $l_1 = 0, 1, 2$, and Path 2 for $l_2 = -2, -1, 0, 1, 2$ in 15 combinations. The OAM of the beam can be identified as follows. Since the mono-mode fiber detector is sensitive only to a Gaussian-type beam ($HG_{00}$) which possesses a single sharp spot (see Figure 5.1), then the OAM requires a hologram with an appropriate helical phase structure design (which varies with $l$) to modify the beam to a Gaussian, and it is the identity of the successful trial-and-error hologram that allows the OAM value to be deduced.[15]

The greatest coincidence rates occur when $l_p = l_1 + l_2$, that is, when OAM is conserved. For example, with $l_p = +1$, the following pairs of OAM produce the highest coincidence rates at the detectors

---

[14]This is a representation of Figure 2 in Mair *et al.* (2001), 'Entanglement of orbital angular momentum states'.

[15]For example, a hologram intended to produce a Gaussian from a beam with $l_2 = -2$ would not allow a detector registration for a beam with $l_2 = +1$.

$(l_1, l_2) = (0, 1), (1, 0), (2, -1)$. While all modes are detected at the end of their path using mono-mode fibers and an avalanche detector, the $l \neq 0$ modes must first be reduced in lateral extent by passing each $l$ through an appropriate hologram. It is noteworthy that out of a singles count rate of $10^5 \, \mathrm{s}^{-1}$, the coincidence rate is $2 \times 10^3 \, \mathrm{s}^{-1}$. The authors ascribe the loss of the second photon in the pair mainly to the transmission losses at the optical surfaces and the efficiency of the detectors. The latter suggestion might well have some reality, for we are reminded of the Grangier experiment in Chapter 3 in which out of $1.8 \times 10^4$ singles counts, only nine coincidences are recorded. This again can be interpreted on the basis that not every photon incident at a detector will register and that certain conditions for absorption must be met.

## 5.5 Skew-Angle Measurement

Leach *et al.* (2006) have succeeded in measuring the skew angle of the Poynting vector for a multi-photon laser-generated helically phased, $\exp(il\phi)$, beam using a Shack Hartmann wavefront sensor. A beam expander takes the output from a HeNe laser and projects it onto a spatial light modulator (SLM) in the form of a computer-generated hologram.[16] This controls the intensity and phase of the light reflected from it, and allows an $LG$ mode to be created. The helically phased beam reflected from the SLM passes through an imaging system that controls the beam width, then into a Shack Hartmann wavefront sensor which leaves an image in a camera behind it.

A Shack Hartmann wavefront sensor comprises an array of lenses such that an incident plane wave results in a corresponding array of spots focused on to a plane of a detector array. Transverse displacement of any of these spots corresponds to a local inclination of the wavefront. (Leach *et al.* 2006, 11919)

---

[16]This is programmed as a $l$-forked diffraction grating; see Figure 5.8 for a single-forked grating hologram.

Both the magnitude and sign of $l$ are discernible with this apparatus. The authors report that a skew angle of 0.1 mrads results in a spot displacement of 7 $\mu$m, with the apparatus being capable of distinguishing displacements as small as 0.7 $\mu$m, corresponding to a change in skew angle of 0.01 mrads.

For the linear momentum density $\varepsilon_o \vec{E} \times \vec{B}$, its components in an OAM beam are found from the wave equation within the paraxial approximation, using a solution in cylindrical coordinates of the form

$$u(r, \phi, z) = u(r, z)\exp(i\phi l) \tag{5.5}$$

For a circularly polarized beam, we have the following (Allen *et al.* 1992; Allen and Padgett 2000):

$$p_r = \varepsilon_o \frac{\omega k r z}{(z^2 + z_R^2)}|u|^2$$

$$p_\phi = \varepsilon_o \frac{\omega l}{r}|u|^2 - \frac{1}{2}\omega\sigma\frac{\partial |u|^2}{\partial r} \tag{5.6}$$

$$p_z = \varepsilon_o \omega k |u|^2$$

where $\varepsilon_o$ is dielectric permittivity, $\omega$ is angular frequency, $k$ is the wavenumber, and $\sigma = +1, -1, 0$ for left-circular, right-circular, and linear polarization, respectively.

However, in Section 2.2.3, it was pointed out that in order to sanction a plane wave to carry SAM, a contradiction would need to be tolerated in that the electric field must vary with the radius, yet it cannot do so if Maxwell's $\nabla \times \vec{H} = -\partial\vec{E}/\partial t$ is to be satisfied. The only consistent conclusion is that the angular momentum spin density (2.12) must vanish, and likewise the second term on the right of (5.6) for $p_\phi$ must also be suppressed. Unfortunately, this leaves us in want of a mechanism by which a single-photon front can carry SAM and OAM. A way out of this dilemma has been suggested in Figure 3.5, where a transverse array of tubes, along each of which a screw thread or HSD advances at speed $c$, is able to transmit SAM to a stationary plane set perpendicular to the propagation axes of the tubes. In fact, the array is capable of transferring SAM to multiple locations, the only limitation being the probability of absorption. However, it still remains to be shown how an array of this kind can convey OAM to the stationary plane, and this will now be addressed.

Leach *et al.* (2006, 11920) maintain that "Beams with helical phase fronts [...] possess an OAM of $l\hbar$ per photon [$l$ is integer]"; in other words, a single photon can carry OAM, which reinforces the suggestion here that a photon has a real extended transverse structure in order that relative phases can exist in a plane perpendicular to the beam. So, it cannot be a point particle and it cannot be a localized wave packet. As we have seen, neither can it be a Maxwellian plane wavefront which, although laterally extended, has no facility to exhibit rotary motion (SAM).

It should be clear from the previous discussion for optical OAM that the linear momentum vector now takes on the skew angle $\gamma = l/kr$ as given by Leach *et al.* (2006), where $\gamma = \tan^{-1}(p_\phi/p_z)$.[17] Here, $r$ is the radius from the OAM beam axis, $l$ is the integer topological charge, and $k$ is the wavenumber of the light. Implicit here are the momentum relationships

$$p_\phi = \frac{l\hbar}{r}, \quad p_z = \frac{\hbar}{\lambda} \tag{5.7}$$

Since the beam is well collimated, there is no radial momentum component and so $p_r$ in (5.6) is taken to be zero.

The principle that the phase (or wavelength) shift is proportional to the thickness of the dispersive medium has been exploited in experiments with spiral phase plates, in which a plane photon front is modified into a LG beam. This is a transparent plate whose thickness varies linearly with the azimuthal angle taken about the plate center. One of the earliest demonstrations of imprinting a spiral wave character onto a TEM$_{00}$ laser beam was reported by Beijersbergen *et al.* (1994) who used a visible wavelength $\lambda = 633\,\mathrm{nm}$.

---

[17]Leach *et al.* (2006, 11920) give the small angle approximation $\gamma = p_\phi/p_z$. In fact, they consider the small angle case $l = 1$, $\lambda = 6.32 \times 10^{-7}\,\mathrm{m}$, and $r = 10^{-3}\,\mathrm{m}$, to yield $\gamma = 6.32 \times 10^{-4}\,\mathrm{rads}$.

Their staircase phase plate was constructed from acrylic (refractive index $n = 1.49$), with the steps radially milled in $5°$ azimuthal increments. Each step was $10\,\mu m$ lower than the previous one, so that a total of 72 steps produced a total descent of $h = 720\,\mu m$. The integer change in helicity (topological charge) is given as $\Delta l = h\Delta n/\lambda$, where $\Delta n$ is the change in refractive index between the plate and its environment. The plate was encased in a brass cell containing a liquid with almost the same refractive index as the acrylic, with $\Delta n = 0.00087$ (which could be varied with temperature), so that $\Delta l \approx 1$.

> When a beam is passed through such a plate, the helical surface can be expected to give a helical character to the beam. A rigorous calculation of the operation of the plate would require vector diffraction theory. However, for beams with small divergence and with a height of step that is sufficiently small we remain in the paraxial regime, so that the operation of the plate can be considered to be an operation on the phase only. (Beijersbergen *et al.* 1994, 322)

The amplitude directly after passage through the plate is given by (5.5) with $l$ replaced by $\Delta l$.

For the moment, let us keep the angular momentum relations (5.7) in mind as we explore the relationship between the momentum and velocity components of the OAM single photon.

## 5.6   Momentum and Velocity Components

Let us consider the propagation and azimuthal directions of a single photon with OAM. Giovannini *et al.* (2015) have altered the transverse spatial structure of a single photon to compare its group velocity with that of a time-correlated plane wave.[18] A pair of photons strongly correlated in wavelength is produced by SPDC. Using a knife-edge prism, the idler and signal are oppositely diverted,

---

[18]They assume that a light photon is a wave packet formed from a bandwidth, but the assumption of a group velocity is not a necessary part of the theory. The notion of a propagation velocity is sufficient.

the former into a beam splitter and the latter through a retardation processor. This consists of a diffractive optical element or spatial light modulator (SLM) that imposes a transverse spatial structure on the signal. This is followed by a second SLM, a known distance $L$, from the first that reverses the change. The signal photon then meets the idler at the beam splitter where coincidence detection equipment is arranged at its two ports.

A calibration can be performed without the SLMs in which a path delay in the signal photon path can be varied until there is a coincidence null, known as a HOM dip. In the classical interpretation, this results from complete constructive interference at one detector and complete destructive at the other, while in the quantum mechanical exposition, both photons are assumed to exit the same port.[19] The SLMs can then be introduced and the path delay adjusted to investigate the consequent shift in the coincidence null. This results from the transit-time delay introduced into the signal photon path between the LSMs.

The Cartesian components $(k_x, k_y, k_z)$ of the wave-vector $k_o = 2\pi/\lambda$ follow the relationship

$$k_o^2 = k_x^2 + k_y^2 + k_z^2 \tag{5.8}$$

The investigators assume a propagation component $k_z$ and a radial component

$$k_r = (k_x^2 + k_y^2)^{1/2} \tag{5.9}$$

However, in the context of OAM in the paraxial limit, (5.9) could just as well be replaced by an azimuthal component

$$k_\phi = (k_x^2 + k_y^2)^{1/2} \tag{5.10}$$

We do not intend here to employ the assumption of a wave packet, and so the group velocity $v_{g,z}$ in Giovannini *et al.* (2015) is to be

---

[19]The authors remark that for emergence to occur from the same port, the arrival times of the two photons at the beam splitter 'are matched to a precision better than their coherence time'.

replaced by a propagation velocity $v_z$, to give

$$v_z = c\left(1 - \frac{k_\phi^2}{k_o^2}\right)^{1/2} \tag{5.11}$$

It follows from (5.8), (5.10), and (5.11) that

$$\frac{v_z}{c} = \frac{k_z}{k_o} = \frac{p_z}{p_o} \tag{5.12}$$

where $p_z$ and $p_o$ are the propagation and skew-angle momenta, respectively. Also, since

$$v_\phi = c\left(1 - \frac{v_z^2}{c^2}\right)^{1/2} \tag{5.13}$$

then from (5.11) and (5.12), we also have

$$\frac{v_z}{v_\phi} = \frac{p_z}{p_\phi} \tag{5.14}$$

For a time delay of $\Delta t\,$s over a distance of $L$ m, the change in the plane wave propagation time due to transverse spatial structuring is given by

$$\Delta t = \frac{L}{c} - \frac{L}{v_z} = \frac{L}{c} - \frac{L}{c}\left(1 - \frac{k_\phi^2}{k_o^2}\right)^{-\frac{1}{2}} \approx -\frac{Lk_\phi^2}{2ck_o^2} \tag{5.15}$$

provided that $k_\phi \ll k_o$. As a hypothetical example, the authors cite the case of $k_\phi/k_o = 4.5 \times 10^{-3}$ and $L = 1$ m to predict a time delay of $\Delta t = 33.8$ fs.[20] In practice, for a Bessel beam, it is reported that for $k_\phi/k_o = 0.00225$ and $k_\phi/k_o = 0.00450$, the spatial delays have been measured to be $c\Delta t = 2.7 \pm 0.8 \times 10^{-6}$ m and $c\Delta t = 7.7 \pm 0.8 \times 10^{-6}$ m, respectively. These can be compared with the theoretical predictions of $c\Delta t = 2.0 \times 10^{-6}$ m and $c\Delta t = 8.1 \times 10^{-6}$ m, respectively. From these values, we can calculate that their SLM separation was $L \sim 0.8$ m, a fact the authors leave unstated.

---

[20]The authors give it as $\sim$30 fs to one significant figure.

The investigators further suggest "a simple geometrical model" to account for the delay, as the transverse spatial structure travels freely in the space between the SLM. This amounts to the plane wave momentum "propagating with an angle $\alpha$ with respect to the optic axis". It is clear from Giovannini *et al.* (2015, Figure 1) and (5.12) that

$$\cos \alpha = \frac{k_z}{k_o} = \frac{p_z}{p_o} = \frac{v_z}{c} \tag{5.16}$$

Taking their results to their unstated conclusion, for a single photon with OAM, the Poynting vector takes a spiral path through space with speed $c$ and momentum $p_o$. This conforms with the model of a photon with OAM given in Figure 5.7. It is presumed that the HSD have a phase that varies with the azimuthal angle around the optic axis. The published literature offers little clarity as to how this is to be represented for circularly polarized rays. A suggestion is shown in Figure 5.9. Figure 5.9(a) exhibits the transverse photon array of HSD in linear motion with no optical OAM and an equal phase for all phasors (arrows at the end of the guide tubes). Figure 5.9(b) shows the Poynting vector, or momentum vector, rotating around the optic axis again with constant phase at all azimuthal angles. This time, the HSD tubes are curved around the optic axis. In order to conserve angular momentum, the HSD at higher radii have their wavelength stretched out (not shown) as the radius from the optic axis increases, thereby decreasing the azimuthal momentum as the radius increases, see Equation (5.7). Figure 5.9(c) adds in the phase delay which varies with azimuthal angle, due to a spatial staggering of the curved guide tubes as might result from a spiral plate. The latter two parts are depicted at a fixed optical OAM radius $r_2$ where $p_\phi = \hbar/r_2$. Again, the Poynting vector, or momentum vector, has two components $p_z$ along the $z$ axis, and an azimuthal component $p_\phi$ rotating around the optic axis.

In Part 2 of this treatise, an SAM mass ring will be set up based on Figure 5.9 in which there is curvature of the HSD array around the optic axis. Since Maxwell's theory will not be used as a basis, there will be no further reference to it.

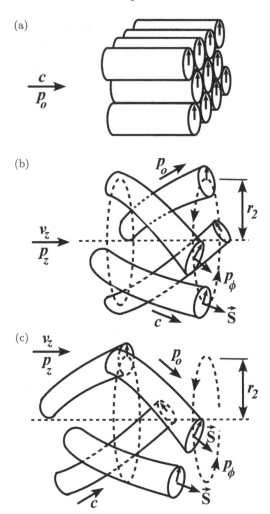

Figure 5.9    The three stages of assembling a single-photon OAM at a particular radius. (a) There is no rotation of the Poynting vector and there are equal phases across the single-photon front. (b) The Poynting vector is now rotated clockwise about the propagation axis and the speed $c$ and momentum $p_o$ reduce to $v_z$ and $p_z$, respectively, along the propagation direction. (c) A phase delay or tube lag is introduced along $z$ that depends on azimuthal angle $\phi$ to produce a spiral transverse structure.

# References

Abramochkin, E., and V. Volostnikov. 'Beam transformations and nontransformed beams'. *Optics Communications*, 83 (1991): 123–135.

Allen, L., M. W. Beijersbergen, R. J. C. Spreeuw, and J. P. Woerdman. 'Orbital angular momentum of light and the transformation of Laguerre–Gaussian laser modes'. *Physical Review A*, 45 (1992): 8185–8189.

Allen L., and M. J. Padgett. 'The Poynting vector in Laguerre–Gaussian beams and the interpretation of their angular momentum density'. *Optics Communications*, 184 (2000): 67–71.

Beijersbergen, M. W., R. P. C. Coerwinkel, M. Kristensen, and J. P. Woerdman. 'Helical wavefront laser beams produced with a spiral phaseplate'. *Optics Communications*, 112 (1994): 321–327.

Courtial, J., and M. J. Padgett. 'Performance of a cylindrical mode converter for producing Laguerre–Gaussian modes'. *Optics Communications*, 159 (1999): 13–18.

Feng, S., and H. G. Winful. 'Physical origin of the Gouy phase'. *Optics Letters*, 26 (2001): 485–487.

Friese, M. E. J., J. Enger, H. Rubinsztein-Dunlop, and N. R. Heckenberg. 'Optical angular-momentum transfer to trapped absorbing particles'. *Physical Review A*, 54 (1996): 1593–1596.

Giovannini, D., J. Romero, V. Potoček, G. Ferenczi, F. Speirits, S. M. Barnett, D. Faccio, and M. J. Padgett. 'Spatially structured photons that travel in free space slower than the speed of light'. *Science*, 347, 6224 (2015): 857–860.

He, H., M. E. J. Friese, N. R. Heckenberg, and H. Rubinsztein-Dunlop. 'Direct observation of transfer of angular momentum to absorptive particles from a laser beam with a phase singularity'. *Physical Review Letters*, 75 (1995): 826–829.

Krenn, M., M. Malik, M. Erhard, and A. Zeilinger. 'Orbital angular momentum of photons and the entanglement of Laguerre–Gaussian modes'. *Philosophical Transactions A: Mathematics, Physics, Engineering, Science*, 375 (2017): 20150442.

Leach, J., M. J. Padgett, S. M. Barnett, S. Franke-Arnold, and J. Courtial. 'Measuring the orbital angular momentum of a single photon'. *Physical Review Letters*, 88 (2002): 257901.

Leach, J., J. Courtial, K. Skeldon, Stephen M. Barnett, S. Frank-Arnold, and Miles J. Padgett. 'Interferometric methods to measure orbital and spin, or the total angular momentum of a single photon'. *Physical Review Letters*, 92 (2004): 013601.

Leach, J., S. Keen, M. J. Padgett, C. Saunter, and G. D. Love. 'Direct measurement of the skew angle of the Poynting vector in a helically phased beam'. *Optics Express*, 14 (2006): 11919–11924.

Mair, A., A. Vaziri, G. Weihs, and A. Zeiliger. 'Entanglement of orbital angular momentum states'. *Nature*, 212 (July 2001): 313–316.

McLaren, M., F. S. Roux, and A. Forbes. 'Realising high-dimensional quantum entanglement with orbital angular momentum'. *South African Journal of Science*, 111(1–2) (2015): 1–9.

Padgett, M., J. Arlt, N. Simpson, and L. Allen. 'An experiment to observe the intensity and phase structure of Laguerre–Gaussian laser modes'. *American Journal of Physics*, 64 (1996): 77–82.

Simon, D. S. 'Quantum sensors: improved optical measurement via specialized quantum states'. *Journal of Sensors*, (2016): Article ID 6051286.

Simpson, N. B., K. Dholakia, L. Allen, and M. J. Padgett. 'Mechanical equivalence of spin and orbital angular momentum of light: an optical spanner'. *Optics Letters*, 22 (1997): 52–54.

Soskin, M. S., V. N. Gorshkov, M. V. Vasnetsov, J. T. Malos, and N. R. Heckenberg. 'Topological charge and angular momentum of light beams carrying optical vortices'. *Physical Review A*, 56 (1997): 4064–4075.

Tang, D. Y., N. R. Heckenberg, and C. O. Weiss. 'Phase dependent helical pattern formation in a laser'. *Optics Communications*, 114 (1995): 95–100.

Wei, H. Q., X. Xue, J. Leach, M. J. Padgett, S. M. Barnett, S. Franke-Arnold, E. Yao, and J. Courtial. 'Simplified measurement of the orbital angular momentum of single photons'. *Optics Communications*, 223 (2003): 117–122.

# Part 2

# Mass Vortex Ring Theory

## Chapter 6

# The Unloaded OAM Mass Ring

*Two types of rings are to be constructed. The first is a spin angular momentum (SAM) mass ring that arises out of trapped helical space dislocations (HSD). The second is an orbital angular momentum (OAM) mass ring, in which the trajectory of the SAM ring is confined to an orbit. The fine-structure constant $\alpha$, upon which the speed of the SAM cluster depends, is central to these definitions. The definitions of mass, charge, and self-potential are incorporated. The concept of spin $1/2$, which quantum mechanics has employed in the construction of the hydrogen atom, will prove to be superfluous in the present model. Coulomb's law is derived from considerations of spin-3 momentum. A modification to dynamics is proposed in which there are two types of acceleration: passive and active. The former corresponds to acceleration without energy absorption and relates to electrostatic attraction and repulsion. The latter involves an exchange of energy with its surroundings.*

## 6.1 Overview

One has to admit that in spite of the concerted efforts of physicists and philosophers, mathematicians and logicians, no final clarification of the concept of mass has been reached. (Jammer 1997, 224)

### 6.1.1 *Electron vortices*

Electron vortices carrying orbital angular momentum (OAM) were first demonstrated by Uchida and Tonomura (2010) by passing a 300-kV beam through a spiral phase plate.[1] Improvements have been

---

[1]Their spiral phase plate was constructed from three slivers of thin graphite film mounted on a copper grid with thicknesses of 10–100 nm. This imprinted a Laguerre–Gaussian (LG) spiral wave character on the plane wave passing

made by Verbeeck *et al.* (2010) by using a computer-designed holographic binary mask to generate doughnut-shaped images with topological charge $l = \pm 1$. By increasing the number of half-slits in the binary mask, McMorran *et al.* (2011) have managed to create topological charge $0, \pm 25, \pm 50, \ldots$, obtained from 25 half-slits at orders $n = 0, 1, 2, \ldots$[2] The authors conclude the following:

> Unlike a classical vortex, this orbital motion cannot be attributed to the collective behaviour of many electrons in the beam; at the low beam currents of this experiment, the separation between individual electrons along the optical axis is several orders of magnitude larger than the longitudinal extent of each wave packet [...] Such high-OAM electron vortex states can exist at rest too, because (unlike a beam of photons) one can produce, using a decelerating electric field, a reference frame in which the forward motion of the electron vortex is zero along the optical axis. (McMorran *et al.* 2011, 194)

Had they taken their work a step further, they might have been able to test for charge switching in a single electron.[3] By entirely reversing the motion of the electron OAM ring passing perpendicularly through a magnetic field, the direction of deflection can be observed and compared with that given by the Lorentz force. Contrary to expectation, an unchanged direction on reversal would show that the sign of the charge has been reversed in the same ring, a prediction of the MVR theory to be developed here.

This chapter sets up the relations for a 'light-ray' vortex theory in the form of an OAM mass ring. The need to begin with an 'unloaded' OAM mass ring, that is, one that contains no absorbed radiation, will become apparent when setting up the Sommerfeld–Dirac fine-structure formula in Chapter 7. It will be seen to be necessary for

---

through it. However, the lack of continuity in the spiral plate resulted in non-integer topological charge.

[2]The relation of topological charge $l$ to the number of half-slits $b$ and order $n$ is $l = nb$.

[3]This is a new concept and serves as a test of the new MVR theory developed here. By this, it is meant its behavior changes from negative to positive charge or vice versa.

the development of the theory of the 'loaded' mass ring, that is, a ring that absorbs radiation, upon which the theory of spectral lines is to be based.

### 6.1.2 Definition of 'spin-2'

A 'particle', such as an electron or proton, is to be constructed by first setting up a spin angular momentum (SAM) mass ring moving linearly in the direction of its axis. This consists of a cluster of helical space dislocations (HSD) confined to helical trajectories, and is identical with the optical OAM that has been observed in the laboratory using cylindrical lenses, spiral wave plates, and holograms (Allen *et al.* 1992; Beijersbergen *et al.* 1994; Friese *et al.* 1996). So, mass has already been created in experiments on optical OAM, but without being fashioned into an OAM mass ring (as we intend to do here) it carries no charge. This angular momentum about its propagation axis, which we denote here as 'spin-2', allows an azimuthal momentum field (or mass field) with a magnitude that varies inversely as the radius from the optic axis.[4] When the linear motion of the SAM mass ring (optical OAM) is brought to rest and the minimum spin-2 radius is considered, the stationary ring provides a definition of 'rest mass' based on the spin-2 perimeter. Here, the Poynting vector has purely azimuthal rotation. The central point is that mass derives its existence from the presence of angular momentum.

To create an OAM mass ring (and consequently charge), the optic axis of the SAM mass ring propagation (optical OAM) is bent into a closed curve. The spin-2 rotation of the HSD momentum vector is to be the basis of magnetic effects (e.g. Lorentz force), whereas spin-3 is shown to have utility in accounting for electrical attraction and repulsion (the Coulomb law). The concept of spin

---

[4]Producing a mass field from a momentum field only requires a division by the constant speed in the field. To be more exact, the mass field is a scalar map which is superimposed upon the velocity field to make a momentum–vector map. This inverse radius dependence is similar to that for optical OAM (Leach *et al.* 2006, 11920), but they are not the same process.

$\pm 1/2$ is superfluous here, and in Chapters 7 and 8, an accurate fine-structure law for hydrogen with Doppler-shift corrections is obtained without it. Finally, Chapter 9 develops these ideas in relation to the hydrogen atom and introduces the hyperfine structure as a result of a line integral of the proton's spin-2 field momentum around the electron's spin-2 azimuthal circuit.[5]

## 6.2   SAM Mass Ring

### 6.2.1   *HSD curvature*

Using only one HSD tube from a single-photon array for illustration, Figures 6.1(a)–6.1(c) show its curvature into a toroid to produce spin-2. The guide tube, which the HSD travels along at speed $c$, has been shown and the surrounding HSD array has been omitted for clarity. A definition of 'rest mass' is possible from the minimum radius $R_{2o}$; see Section 6.2.2.

The HSD is now set in motion along the $z$ axis at speed $v_z = \alpha c$, by slicing through the spin-2 guide tube cross-section (in a half-plane containing the $z$ axis) and stretching it out, as shown in Figure 6.1(d). Here, $\alpha$ is the fine-structure constant. This tube can be iterated periodically along the $z$ axis as the SAM mass ring progresses along it, and in Figure 6.1(d), one such tube cycle is shown. The utility of this choice for $v_z$ will become apparent in the subsequent derivation of the Sommerfeld–Dirac fine-structure formula for hydrogen in Chapter 7. Consequently, we must have for all SAM rings

$$v_z = \alpha c$$
$$v_2 = c(1 - \alpha^2)^{1/2} \tag{6.1}$$

Equations (6.1) ensure that $v_2^2 + v_z^2 = c^2$. We note the difference between a spin-1 and spin-2 helix. In Figure 6.1(a), for spin-1, the entire HSD (circular polarization screw thread) is in linear motion at speed $c$ along the $z$ axis, without rotation, through a guide tube that merely assists visualization. For example, the right-wound thread

---

[5]This circuit will be elliptical with radial and azimuthal (spin-2) components.

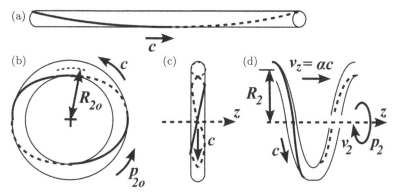

Figure 6.1 Construction of a SAM mass ring with attention confined to only one tube from the single-photon HSD array. (a) A single spin-1 (circular polarization) guide tube along which a screw thread or HSD travels linearly with speed $c$. (b) The tube is bent into a torus of radius $R_{2o}$ with phase-matched ends to construct a rest mass with spin-2. (c) Left-end elevation of (b). (d) The toroidal SAM mass ring in (c) is cut through its cross-section and stretched out. The thread follows the iterated helical guide tube and moves at speed $\alpha c$ along its optic axis $z$, about which it rotates at speed $v_2 = c(1 - \alpha^2)^{1/2}$, where $\alpha$ is the fine-structure constant.

shown in Figure 6.1(a) makes available a counter-clockwise motion or left-circular polarization in a stationary plane set perpendicular to its axis through which it passes. However, in Figure 6.1(d), for spin-2, the helically wound tube marks out the trajectory that the spin-1 thread follows at speed $c$. For example, the right-wound helix shown in Figure 6.1(d) allows a spin-1 momentum to rotate clockwise and impart a spin-2 clockwise rotation to an object it encounters; see also Figure 5.7. As stated, this construction of an SAM mass ring is equivalent to optical OAM.

### 6.2.2 *Spin-2 rest mass*

Using only one HSD tube from a single-photon array for illustration, we posit here that each tube of the transverse HSD array — only one of which is shown in Figure 6.1 — preserves angular momentum $\hbar$ about the spin-2 axis. Since optical OAM derives from redirected linear momentum in the HSD, the sign of the SAM mass ring rotation is independent of the circular polarization sense.[6] First, we define a spin-2 momentum at the minimum spin-2 radius $R_{2o}$ for the ring at rest, as shown in Figures 6.1(b) and 6.1(c),

$$p_{2o} = \frac{h}{2\pi R_{2o}} \tag{6.2}$$

From (6.2), we define the spin-2 rest mass as

$$m_{2o} = \frac{h}{2\pi R_{2o}c} \tag{6.3}$$

In (6.3), it is the length $2\pi R_{2o}$ that defines the mass. For the present, we shall assume this perimeter of the spin-2 rotation to be circular, but in dealing with the hyperfine structure in Chapter 9, elliptical perimeters will be introduced. We further posit that the SAM ring at rest has a spin-2 momentum vector $\vec{p} = m_{2o}\vec{c}$ that runs along the varying direction $\hat{c}$ of the HSD (spin-1) tube. In consequence, the calculation of the SAM ring energy, when the ring is at rest, is not to be obtained from the relativistic four-vector $(E/c, p)$ but from a scalar product of the SAM ring momentum $\vec{p} = (m_{2o}c(1 - \alpha^2)^{1/2}, m_{2o}\alpha c)$ with the velocity vector $\vec{c} = (c(1 - \alpha^2)^{1/2}, \alpha c)$ partitioned into azimuthal (spin-2) and linear components; thus,

$$E = \vec{p} \cdot \vec{c} = m_{2o}c(1 - \alpha^2)^{\frac{1}{2}} \cdot c(1 - \alpha^2)^{\frac{1}{2}} + m_{2o}\alpha c \cdot \alpha c = m_{2o}c \cdot c \tag{6.4}$$

When the OAM ring is in motion, the $\vec{p}$ and $\vec{c}$ components will be in need of modification. Division by $(1 - \alpha^2)$ yields

$$m_{2u}c^2 = m_{2o}c^2 + m_{2u}(\alpha c)^2 \tag{6.5}$$

where

$$m_{2u} = m_{2o}(1 - \alpha^2)^{-1} \tag{6.6}$$

---

[6]While the sign of the optical SAM from circular polarization depends on the induced rotation in a plane — left-circular polarization is $+\hbar$, while right-circular polarization is $-\hbar$ — its linear momentum is the same for both cases. Optical OAM derives from linear momentum redirected into azimuthal rotation.

Here, $m_{2u}$ is to be denoted as the 'unloaded mass', which means that no radiation has been absorbed.[7] From the conservation of angular momentum, we can modify (6.3) to give

$$m_{2u} = \frac{h}{2\pi R_2 v_2} = \frac{h}{2\pi R_2 c (1 - \alpha^2)^{1/2}} \tag{6.7}$$

where $R_2$ is the spin-2 radius for an SAM ring in linear uniform motion; see Figure 6.1(d).

It follows from (6.3), (6.6), and (6.7) that the lower-bound radius $R_2$ — see Figure 6.1(d) — for the SAM mass ring in motion is related to the rest mass radius $R_{2o}$ by[8]

$$R_2 = R_{2o}(1 - \alpha^2)^{1/2} \tag{6.8}$$

### 6.2.3 *Spin-2 field mass*

A more detailed representation of the SAM mass ring with the surrounding HSD array tubes included is shown in Figure 6.2 at the minimum spin-2 radius $R_2$. In order to define a field mass, we must first define a field momentum. This is the momentum in the space surrounding the base radius spin-2 rotation.

$$p_{2f} = m_{2f} v_2 = \frac{\hbar}{R_{2f}} \tag{6.9}$$

where $m_{2f}$ is the field mass and $R_{2f}$ is the radius from the spin-2 axis.[9] In Figure 6.2(a), we see the front elevation of the SAM mass ring in its form as a cluster of nested HSD following helical trajectories around the $z$ axis at speed $c$. For clarity, this view has been stretched horizontally, and those HSD at higher radii have been omitted.[10] Figure 6.2(b) shows the right-end elevation of

---

[7]Principle (7.14) in Clarke (2017, 253) does not anticipate an unloaded ring, so gives (6.6) as $m_{2u} = m_{2o}(1 - \alpha^2)^{-1/2}$.

[8]The unloaded radius will be denoted by the uppercase $R$ and the loaded (see Chapter 7) by the lowercase $r$.

[9]This is given in Principle 7.5 (Clarke 2017, 236), but the distinction between loaded and unloaded OAM rings had not then been developed.

[10]In fact, the angle that a tube makes with the propagation axis $z$ is $\cos^{-1}(\alpha)$, where $\alpha$ is the fine-structure constant.

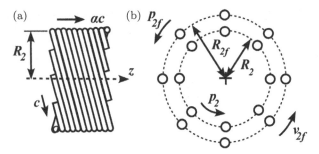

Figure 6.2   Detailed view of an SAM mass ring traveling at speed $v_z = \alpha c$ with spin-2 angular momentum $\hbar = p_2 R_2$ at its lower-bound radius. (a) Front elevation of nested HSD at lower-bound radius $R_2$. (b) Right-end elevation of (a) showing the lower-bound spin-2 radius $R_2$. Part of the surrounding HSD field is shown at a radius $R_{2f} > R_2$ which also follows helical trajectories at speed $c$, and possesses angular momentum $\hbar = p_{2f} R_{2f}$.

Figure 6.2(a). Other tubes in the array follow helical trajectories at a radius $R_{2f} > R_2$ with field momenta $p_{2f} < p_2$ while preserving spin-2 angular momentum $\hbar$.

The spin-2 field mass $m_{2f}$ can be defined as a modification of (6.9), by dividing the field-momentum $p_{2f}$ by the spin-2 field speed $v_{2f} = c(1-\alpha^2)^{1/2}$ which is a constant of the field, that is, independent of $R_{2f}$; thus,

$$m_{2f} = \frac{\hbar}{R_{2f}c(1 - \alpha^2)^{1/2}} \qquad (6.10)$$

This defines a scalar map of the mass field and it depends on the radius $R_{2f}$ from the $z$ axis. A diminution of field mass results from a stretching out of the azimuthal HSD components, an effect that increases with radius $R_{2f}$.

## 6.3   OAM Mass Ring

### 6.3.1   *Definition of 'spin-3'*

We now proceed to set up an OAM mass ring from the SAM cluster of HSD. Let us first define a 'spin-3' rotation as that which occurs around the axis of the OAM mass ring. The $z$ axis along which the

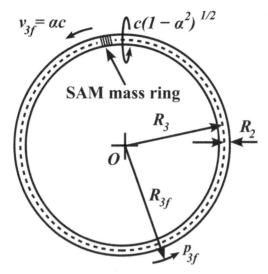

$v_{3f} = \alpha c$     $c(1 - \alpha^2)^{1/2}$

**SAM mass ring**

$R_3$    $R_2$

$O$

$R_{3f}$

$p_{3f}$

Figure 6.3   An unloaded OAM mass ring with radius $R_3$, which represents the closed-circle trajectory that a single SAM mass ring follows. As the HSD follow their periodically iterated helical guide tubes in the SAM mass ring at speed $c$, the SAM cluster progresses around the OAM ring at speed $\alpha c$.

SAM ring moves in Figure 6.1(d), or more generally in Figure 6.2(a), is now curved into a toroid to form an OAM mass ring.[11] This is shown in Figure 6.3, with constant radius $R_3$, and an azimuthal speed $v_{3f} = \alpha c$ which is constant at all spin-3 field radii $R_{3f}$, where on the OAM ring we have $R_3 - R_2 \leq R_{3f} \leq R_3 + R_2$. This means that the spin-3 angular speed must vary with $R_3$ on the ring to maintain a constant $v_3 = \alpha c$. The HSD pass along the guide-tube trajectories in the SAM mass ring shown in Figure 6.2(a) at speed $c$, and progress along the OAM mass ring circumference at speed $v_3$. The spin-3 field momentum is shown as $p_{3f}$.

---

[11]In Principle 7.11 in Clarke (2017, 247), it was thought that SAM rings were linked longitudinally around the OAM ring. It is now proposed that they are only one rotation or wavelength in extent, and that they follow heuristic screw-thread guide tubes that are longitudinally iterated. Figure 7.12 in that work is therefore in error.

### 6.3.2 *Spin-3 field momentum*

Generalizing (6.1), the corresponding field velocity components are independent of the field radius $R_{3f}$ and are

$$v_2 = v_{2f} = c(1 - \alpha^2)^{1/2}$$
$$v_3 = v_{3f} = \alpha c \tag{6.11}$$

The corresponding momentum field is the product of (6.10) and (6.11); thus,

$$p_{3f} = m_{2f}v_3 = \frac{\hbar \alpha}{R_{2f}(1 - \alpha^2)^{1/2}} \tag{6.12}$$

### 6.3.3 *Resultant field momentum*

The spin-2 field speed $v_{2f}$ is tangential to the spin-2 rotation around the $z$ axis; see Figure 6.2(b). The spin-3 field speed $v_{3f}$ acts tangentially to the curved $z$ axis perpendicular to $v_{2f}$. Both are to be an invariant of the field. Since the two components are always mutually perpendicular, this allows the speed along the stretched out HSD in the field to remain constant at $c$; thus,

$$v_{2f}^2 + v_{3f}^2 = (c(1 - \alpha^2)^{1/2})^2 + (\alpha c)^2 = c^2 \tag{6.13}$$

If we multiply this by the square of (6.10) then after recourse to (6.9), (6.10), and (6.12), we have

$$p_{2f}^2 + p_{3f}^2 = \frac{\hbar^2}{R_{2f}^2(1 - \alpha^2)} = m_{2f}^2 c^2 \tag{6.14}$$

Although there is a conservation of spin-2 angular momentum in the field, (6.12) informs us that due to its dependence on $R_{2f}$ there is no such conservation for spin-3, because its radius of rotation $R_{3f}$ is taken from the OAM ring center.

### 6.3.4 *Momentum and radial relations*

We postulate that for the unloaded OAM ring, the spin-3 angular momentum *on the ring* mirrors that of spin-2 in (6.9) with $R_{2f} = R_2$, so that

$$\bar{p}_3 = \frac{Z\hbar}{\bar{R}_3} \qquad (6.15)$$

Although we use $Z$ here, Padgett and Allen (2000, 275, 268) refer to "$l$ intertwined wavefronts" and Padgett *et al.*, (2004, Figure 1 caption) comment on "the number of $l$ intertwined helical phase fronts". Equation (6.15) represents the average momentum $\bar{p}_3$ at the average spin-3 radius $\bar{R}_3$ on the ring. The momentum $p_3$ will vary over the spin-3 ring radius $R_{3f}$, where $R_3 - R_2 \leq R_{3f} \leq R_3 + R_2$. So, we need to calculate an average radius $\bar{R}_3$ at which there is an average momentum $\bar{p}_3$, and this is carried out in Appendix A. Noting that $p_2 = m_{2u}c(1-\alpha^2)^{1/2}$, let us define $\bar{p}_3 = m_{2u}\alpha c$; then, from (6.15) and (A.3), we have for $Z = 1$

$$\frac{\bar{p}_3}{p_2} = \frac{R_2}{\bar{R}_3} = \frac{\alpha}{(1-\alpha^2)^{1/2}} \qquad (6.16)$$

and from (6.6), (6.15), and (6.16), we have

$$R_2 = \frac{\hbar(1-\alpha^2)^{1/2}}{m_{2o}c} \qquad (6.17)$$

having used $\bar{p}_3 = m_{2u}\alpha c$.

### 6.3.5 *Spin-3 energy*

For the calculation of energies, we focus on a single electron ring. As stated, an OAM ring that has absorbed no radiation is to be denoted an 'unloaded OAM mass ring'.[12] This is in contrast to a 'loaded OAM mass ring' (see Chapter 7) which has absorbed radiation and has a

---

[12]This concept is implied in Clarke (2017). A ring should be able to exist irrespective of whether or not it has absorbed radiation and so a radiation-free structure must be available. An SAM mass linear trajectory passing through a very strong electric momentum field (spin-3 $B$ field) rotating perpendicularly to it adopts the same rotation sense. In Chapter 8, we shall see that the OAM ring of the bound state electron unravels as it passes over the proton OAM ring. This can only be because the contrary spin-3 momentum rotation of the proton is countering or nullifying the electron's angular momentum sense.

choice of energy levels to adopt without the need of an external electric potential to form them (McMorran *et al.* 2011, 194).

The problem now is to decide on the treatment of the momentum running along the $z$ axis in Figure 6.1(d) when the trajectory is curved into an OAM mass ring as spin-3 rotation. There are several points to make at this juncture.

(i) The spin-3 momentum $p_3$ varies over $R_{3f}$, where $R_3 - R_2 \leq R_{3f} \leq R_3 + R_2$, and so we shall take the average radius $\bar{R}_3$ — computed in Appendix A — and equate the average spin-3 mass $\bar{m}_{3u}$ at this radius to the spin-2 unloaded mass so that $\bar{m}_{3u} = m_{2u} = m$.[13]

(ii) The spin-3 rest mass, which only exists for the purpose of calculation, we take to be $m_{3o} = m_{2o} = m_o$, from (6.3).

(iii) The spin-3 energy follows (6.5) so that with the identities (i) and (ii) and $\hbar = \bar{m}_{3u} \alpha c \bar{R}_3$, we have for the electron energy

$$mc^2 = m_o c^2 + \frac{\hbar \alpha c}{\bar{R}_3} \qquad (6.18)$$

This equation will be crucial as a starting point for the derivation of the Sommerfeld–Dirac fine-structure formula in Chapter 7 from the new MVR theory.

The fine-structure formula is usually derived by including a potential energy term in the electron energy. In fact, if we replace $\alpha$ with $Z'\alpha$ (traditionally, $Z'$ is the atomic number of the field source) in the second term on the right of (6.18), an accurate fine-structure formula can be obtained.[14] However, there are two objections to this maneuver.

(1) The electron OAM mass ring does not need an external potential in order to possess energy levels:

Electrons can be prepared in quantized orbital states with large OAM, in free space devoid of any central potential,

---

[13]The convention will be adopted here that all unprimed variables refer to an electron, while primed ones denote the proton.

[14]This can be seen in Chapter 7 if we replace $\alpha \to Z'\alpha$ in (7.20).

electromagnetic field, or medium that confines the orbits. (McMorran *et al.* 2011, 194)

(2) Since this term in (6.18) is a consequence of spin-3 electron motion, it has no connection to an external source. Instead, we shall understand it to be a self-potential, an addition to the electron rest mass in consequence of its spin-3 curvature. In fact, the smaller the radius $\bar{R}_3$, the greater is its contribution to the mass $m$.

Figure 6.4 shows an OAM mass ring parameterized by the angles $\theta$ and $\gamma$. The coordinates of point A on the spin-2 circuit are given by the unloaded OAM mass ring position vector in Cartesian coordinates

$$\vec{R}_{3f} = \begin{pmatrix} R_2 \cos\gamma \\ (R_3 + R_2 \sin\gamma)\sin\theta \\ (R_3 + R_2 \sin\gamma)\cos\theta \end{pmatrix} \qquad (6.19)$$

Here, the spin-3 radius $R_{3f}$ is not confined to the $y-z$ plane, but extends to any point on the spin-2 ring.

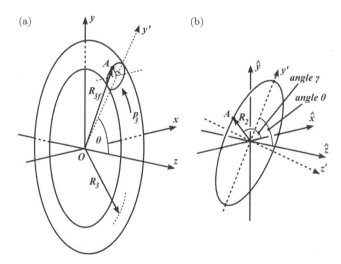

Figure 6.4   (a) Parameterization of point A at radius $R_{3f}$ on the SAM mass ring revolving around an OAM mass ring, with the latter ring normal lying along the $x$ axis. (b) Expanded view of the cross-section $A$ in part (a), where the $y'$ and $z'$ axes lie in the $y-z$ plane.

### 6.3.6    *The momentum-field map*

Figure 6.5 shows the map of the momentum field for an OAM mass ring.[15] While this might be proposed as the structure of all 'particles', for the present purpose, we confine our attention to the proton and electron. The difference between them is only a matter of scale. For example, the definition of the proton mass can follow (6.7) as

$$m'_{2u} = \frac{\hbar}{R'_2 v_2} = \frac{\hbar}{R'_2 c(1 - \alpha^2)^{1/2}} \tag{6.20}$$

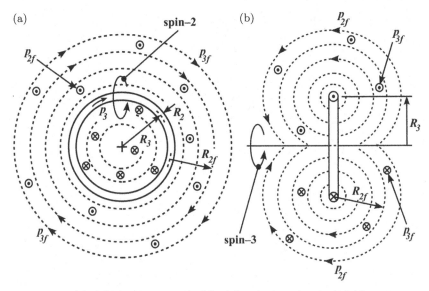

Figure 6.5    (a) OAM mass ring model with spin-2 and spin-3 field-momentum components $p_{2f}$ and $p_{3f}$, and radius $R_{2f}$ for spin-2 field rotation. (b) Right-end elevation of part (a). The dotted and crossed circles are field-momentum components running out of and into the page, respectively. The magnitudes of the $p_{2f}$ momentum components on any given dashed circle are equal. The spin-2 magnetic momentum and spin-3 electric momentum-field components are mutually perpendicular.

---

[15]This is an improved version of Figure 9.1 in Clarke (2017, 303). Part (a) of that figure has incorrect crosses which should only have appeared within the confines of a co-axial cylinder passing perpendicular to the page through the ring aperture.

so that replacing the primed variables with non-primed ones then using (6.16) and (6.17), we have for the proton–electron rest mass ratio a variation of (A.15)

$$\frac{m'_{2o}}{m_{2o}} = \frac{m'_o}{m_o} = \frac{R_2}{R'_2} = \frac{\bar{R}_3}{\bar{R}'_3} \qquad (6.21)$$

Referring to Figure 6.5, there are only two possible OAM rings: the one shown, and another in which the spin-2 rotation takes on the opposite sense while the spin-3 rotation stays the same. As stated earlier, the linear momentum of a HSD is the same for both left- and right-circularly polarized rays. It is the redirection of this linear momentum into azimuthal rotation that creates the SAM mass ring. The rotation of the SAM trajectory into spin-3 rotation produces the OAM mass ring.

Neither of these two OAM ring configurations is to be identified with a positive or negative charge. Instead, we shall look at the motion of these two ring possibilities in electric and magnetic fields, and enquire as to the conditions under which positive and negative charge behavior is exhibited. This is a crucial point: electric charge is to be defined as a *behavior* of the ring in electric and magnetic fields, not as a *property* of it. This behavior depends on the orientation of the ring, that is, whether the spin-3 axis is parallel or antiparallel to an external electric field. We note that $p_{2f}$ carries the magnetic momentum and $p_{3f}$ the electric momentum in the field, and later we shall see that the magnetic field vector $\vec{B}$ follows the direction of $\vec{p}_{2f}$, while the direction of the electric field vector $\vec{E}$ lies along the OAM mass ring axis.

## 6.4 Ring Interactions

### 6.4.1 *Positive and negative charge behavior*

#### 6.4.1.1 *Electrical effects*

According to the model developed here, an electrical effect is a spin-3 effect and is a relationship between rotation senses; see Figure 6.6. This is in contrast to magnetic effects which rely on a changing

**spin–2 momentum field**

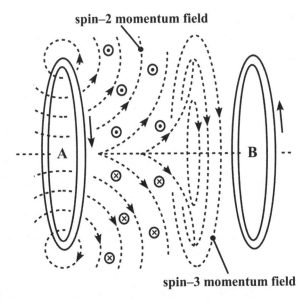

**spin–3 momentum field**

Figure 6.6   The spin-3 momentum field from source OAM mass ring A in the locality of target OAM mass ring B. The source and target rings are shown with aligned axes and opposite spin-3 rotations so they electrically attract. The dotted and crossed rings show spin-3 momentum passing into and out of the page, respectively. The momentum field from ring B that affects A exists but is not shown.

spin-2 source field.[16] As we shall see, both in this section and in the *Magnetic effects* section that follows, there are attraction–repulsion experiments that could be set up that might falsify accepted ideas and favor the new MVR theory.

In judging which pairs of OAM rings might electrically attract and which might repel, consider the following. Allow a counterclockwise ring and a clockwise ring, as viewed in their common direction of motion, to approach a positively charged plate. With spin-3 rotation carrying electrical momentum, one should be attracted to the plate and the other repelled. This means the two rings are exhibiting opposite charge behavior. Let us now dispense with the charged plate and reposition one ring behind the other on a common axis as viewed

---

[16]This is explored in more detail in Chapter 9 when setting out the hyperfine theory.

in their common direction of motion. Having been shown to exhibit an opposite charge effect, we should expect them to attract each other and so we can associate opposite-sense rotations with attraction and same-sense ones with repulsion. We posit here that the axes of the rings must be in alignment for electrostatic interaction to occur,[17] and that if the axis of one runs outside the aperture of the other there is no electrostatic effect.

Now, we cannot assign a charge sign (positive or negative) to a particular OAM ring. If we did, then because the rotation sense depends on which side of the ring one is positioned, the charge sign would depend on observer location. Let us examine this further. In Figure 6.7(a), a length of copper wire is provided with a potential difference between its ends. This leaves a positive dominance at one end of the wire P and a negative one at the other end Q. Allow also that when a potential difference is maintained, the axes of the OAM rings are in alignment, so that viewed toward the left, at location P on the wire, the positive charges have a counterclockwise rotation and at Q the negative charges take on a clockwise rotation. We now bend the two ends of the wire into the page to make a loop with a gap between the ends P and Q; see Figure 6.7(b). Let us now place an OAM ring R in the gap with its axis aligned with P and Q. When viewed toward the left, R has a counterclockwise rotation. This is in the opposite sense to P but the same as Q, so it is attracted toward P which we posited to be positively charged. So, R exhibits negative charge behavior.

Reversing the polarity of P and Q with a switch results in a reversed displacement of the positive and negative rotations in the wire; see Figure 6.7(c). As expected, location P on the wire is now negative and Q is positive. The ring R is now attracted to Q since it has the opposite-sense rotation. So far, the MVR theory is giving the expected result. However, let us return to Figure 6.7(b). Instead of reversing the polarity, let us switch the field off so that the ring R is not reorientated by the manoeuver we are about to perform. Now, rotate the wire about an axis perpendicular to the page so that

---

[17]This point was made in Clarke (2017, 279).

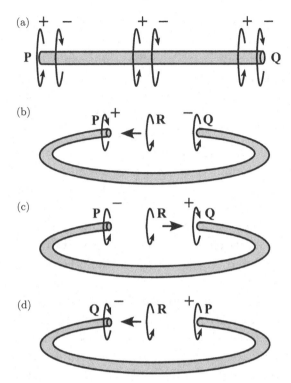

Figure 6.7    Demonstration of charge-switching behavior in an electric field. (a) A potential difference displaces positive and negative rotations in a straight wire. (b) The wire ends are bent round into the page and the OAM ring R is placed between ends P and Q. (c) The potential difference is reversed and the rotation senses at P and Q reverse. (d) Instead of (c), the potential difference in (b) is switched off, the wire rotated about an axis perpendicular to page, then potential difference switched on.

P and Q are juxtaposed, see Figure 6.7(d).[18] The switch is now closed so that the same potential difference is restored to ends P and Q as in Figure 6.7(b). Viewed toward the left, Q which has the negative charge is rotating clockwise, while P which has the positive charge has a counterclockwise rotation. Since R is still counterclockwise, it should be attracted to the opposite-sense rotation Q. However, R has displayed negative charge behavior in Figure 6.7(b), but in

---

[18]Turn the page upside down without turning R.

being attracted to negative Q it now shows positive charge behavior in Figure 6.7(d). The same ring R can behave as if it had positive and negative charge. To test this, it should not be difficult to set up an experiment with cathode rays (OAM rings with identical rotation senses) passing through the apparatus having a direction of motion that is slightly out of alignment with the line joining P and Q. Deflections could then be observed.

The effect described in changing from Figure 6.7(b) to Figure 6.7(d) could just as well be obtained by rotating the axis of ring R in Figure 6.7(b) by $\pi$ rads. It would then be attracted to Q which initially had the opposite rotation sense. The point that emerges here is that charge sign (+ or −) has no meaning without a field for comparison. The sign of the charge can change if the axis of an OAM ring is reversed *in relation to* an existing field.

### 6.4.1.2 *Magnetic effects*

In the new MVR model, magnetic effects result from spatial or temporal variation of spin-2 interactions. There are two ways this can occur. A single-source ring is either thrust toward or away from a target ring so that its spin-2 magnetic momentum-field cutting the target ring's spin-2 circuit increases or decreases. This idea will be developed further in Chapter 9 when discussing the hyperfine interaction. Alternatively, a target ring can pass through a constant magnetic field formed from multiple sources. In this case, the renewal of the spin-2 interaction strength arises from the moving ring's encounter with field momentum from changing or different sources as it progresses.[19]

Let us consider an experiment involving the Lorentz force which can be set up in such a way as to distinguish between present ideas of charge deflection and those connected with the MVR theory. Let us send a stream of cathode rays through an evacuated tube such that the velocity runs perpendicularly through an external magnetic

---

[19]This was the idea behind the treatment of the Lorentz force in Clarke (2017, 272), Equation (8.4).

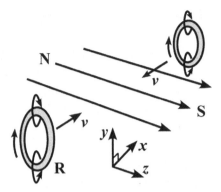

Figure 6.8    Demonstration of charge-switching behavior in a magnetic field. OAM ring passing perpendicularly between a north (N) and south (S) pole of a magnet at speed $v$.

field $B$. Figure 6.8 shows a single electron OAM ring R at the bottom left moving with speed $v$ in the positive $x$ direction. The direction of rotation of its spin-2 momentum is shown at the top and bottom of the ring. The rays will have already been identified as showing negative charge behavior in consequence of their travel from cathode to anode. According to the Lorentz force law, we have $\vec{F} = q\vec{v} \times \vec{B}$, so that with $q$ negative, in Figure 6.8, the ring R is deflected in the positive $y$ direction.

However, let us now reverse the direction of motion of the cathode rays and ring R with an electric field strong enough to arrest the motion. In Figure 6.8, we now move the ring from its top right position to run in the negative $x$ direction. With the change of sign for $v$, the Lorentz force now predicts that the deflection should now be in the negative $y$ direction. However, according to the MVR theory, nothing about the relationship between the external field and the spin-2 rotation sense for R has changed; the deflection should still be in the positive $y$ direction. The reason for this is that it makes no difference whether the ring R travels in the positive or negative $x$ direction through the field. The relative spin-2 senses of the magnetic field sources and ring R remain the same, and the same number of different field sources is encountered per second. If we choose to reverse the direction of R in the middle of the field, then we could claim that the field strength is constant and neither increases nor decreases. The consequence of this is that on reversing the velocity direction, the tendency of ring R to keep deflecting in the positive $y$ direction means that it now exhibits positive charge behavior, that is, it has charge-switched. A strong electric field could be activated to arrest the motion of R, as it moves from its position bottom left, and reverse it so the experiment is feasible.

It is a fact though that cathode rays emitted and accelerated in the negative $x$ direction are deflected in the negative $y$ direction. They follow the Lorentz force law. So, what is the explanation for the difference in the two cases outlined here?[20] Let us consider cathode rays expelled from a heated source and accelerated to run in the negative $x$ direction. Their OAM ring axes are parallel to that of ring R, but they have been rotated through $\pi$ rads. Consequently, the ring's spin-2 rotation senses are reversed. This reversal does not occur for an emission in the positive $x$ direction if the orientation of the ring axis is undisturbed and the direction of motion is reversed by a strong electric field. It is an experiment that could be set up without too much difficulty and would decide between present expectation and the new MVR theory.

## 6.4.2 *Coulomb's law*

### 6.4.2.1 *Effect of proton field on electron ring*

We now discuss the effect of the field from a source ring inhabiting a target ring. When two rings A and B with opposite rotation senses and aligned axes interact, the spin-3 momentum field from A displaces spin-3 momentum from ring B into linear motion so that B moves toward A; see Figure 6.6. So, the proton field energy occupying the electron ring displaces electron spin-3 momentum into linear motion along the direction of its axis. This effect is reciprocal. The electron field energy occupying the proton ring displaces some

---

[20]One case is reversing the direction of motion of ring R that was initially emitted in the positive $x$ direction, and the other case is the emission of ring R in the negative $x$ direction.

of the resident proton energy into linear energy of proton motion. In contrast, when the rotations are in the same sense, the displaced OAM ring momentum produces a linear motion of the ring that is oppositely directed to the attraction case. To quantify this process and obtain a rationale for the direction of motion of B, we need to derive Coulomb's law of electrostatic interaction.

In order to obtain the law of conservation of momentum, we need to argue for the invariance of the field effect of one ring on another when one, or both, is in motion. In other words, the spin-2 field momentum surrounding a ring does not depend on how much of the resident ring energy has been displaced. The linear speed of the electron ring increases as it approaches the proton coaxially, and so the electron SAM trajectory leaves the ring plane and takes on a helical trajectory with increasing rake angle. This diminishes the resident electron energy in the ring plane and apparently the field (which originates from the ring plane) that it can generate to affect the proton ring. However, we shall posit here that the spin-2 axis on the helical trajectory does not become aligned with that trajectory but remains parallel to the host's ring plane. This allows both the field it generates and the field it absorbs to remain constant. So, the proton and electron maintain equal field effects on each other as they approach each other coaxially outside the oscillation boundary.

Consider the interaction between a proton and an electron on a common axis; see Figure 6.9(a). As expressed in (6.21), and referring to (6.6), the ratio of the proton and electron spin-3 radii is the same as that of their rest masses. Following (6.9), we have the spin-2 proton field momentum

$$p'_{2f} = \frac{\hbar}{R'_{2f}} \tag{6.22}$$

Equation (6.16) also holds for the ratio of the *field momenta*, so the spin-3 momentum-field vector for $Z'$ protons in the source field — $Z'$ units of spin-2 angular momentum — is

$$\vec{p}'_{3f} = Z'p'_{2f}\frac{\alpha}{(1-\alpha^2)^{1/2}} \begin{pmatrix} 0 \\ -\cos\theta \\ \sin\theta \end{pmatrix} \tag{6.23}$$

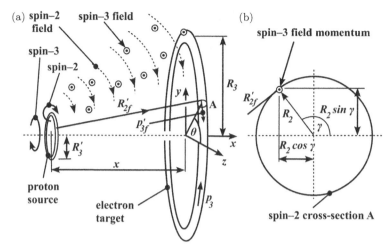

Figure 6.9   (a) Spin-3 field-momentum $p'_{3f}$ from a proton ring source (left) with spin-2 field radius $R'_{2f}$, affecting an electron OAM mass ring (right) with spin-3 radius $R_3$. At the end of the proton's spin-2 radius $R'_{2f}$ is the spin-2 field-momentum $p'_{2f}$ which falls off inversely as the radius. It is perpendicular to the field radius and tangential to the spin-2 field lines of rotation. The spin-3 field-momentum $p'_{3f}$ (which acts out of the page above the ring axis as a dotted circle) is perpendicular to $p'_{2f}$ and is tangential to the electron OAM ring. (b) Expanded view of spin-2 cross-section A shown in part (a). The electron spin-2 radius is $R_2$. The spin-3 field-momentum $p'_{3f}$ at the end of this radius acts out of the page (dotted circle).

However, the extent of the spin-3 field action conveyed to the SAM HSD cluster in the electron ring depends on the perimeter of the spin-2 circuit, which for the present we shall assume to be a circle.[21] So, we need to consider the component of the spin-3 field momentum passing perpendicularly through the target spin-2 perimeter, and take the circuit vector on the electron ring as

$$
d\vec{R}_2 = R_2 d\gamma \begin{pmatrix} 0 \\ \cos\theta \\ -\sin\theta \end{pmatrix} \tag{6.24}
$$

---

[21]In Chapter 9, when discussing the hyperfine effect in the hydrogen atom, the spin-2 perimeter will remain constant while taking on an elliptical profile.

For the case of opposing rotations considered here, this is oppositely directed to $\vec{p}'_{3f}$. Although the increment $dR_2$ lies on the spin-2 perimeter, the circuit vector is to be taken perpendicular to it and parallel to the electron spin-2 axis.

We also need to take account of how the presence of multiple masses in the target ring, indicated by the $Z$ in (6.15), affects the spin-2 integration range of that ring. With this in view, we posit that the spin-2 target-ring angle should be taken over the range $0 \leq \gamma \leq 2\pi$, with the interpretation that there are $Z$ SAM mass rings in one circuit. So, the total spin-3 field action transmitted to the electron SAM cluster through its spin-2 contour is given by

$$J'_{3f} = Z \int_{\gamma=0}^{\gamma=2\pi} \vec{p}'_{3f} \cdot d\vec{R}_2 \tag{6.25}$$

Using (6.22), (6.23), and (6.24), and assuming negligible variation of $R'_{2f}$ with $\gamma$, we get

$$J'_{3f} = -\frac{Z'Z\hbar\alpha}{R'_{2f}(1-\alpha^2)^{1/2}} 2\pi R_2 \tag{6.26}$$

Dividing the action $J'_{3f}$ by the electron spin-2 time period $T_2$ which is obtained from (6.1) with $2\pi R_2/T_2 = c(1-\alpha^2)^{1/2}$ for an OAM ring at rest, we find the proton field energy occupying the electron ring

$$E'_{3f} = -\frac{Z'Z\hbar\alpha c}{R'_{2f}} \tag{6.27}$$

All that remains is to insert $\hbar\alpha c = e^2/4\pi\epsilon_o$ and we arrive at

$$E'_{3f} = -\frac{Z'Ze^2}{4\pi\epsilon_o R'_{2f}} \tag{6.28}$$

which is Coulomb's law. If we use the spin-3 radius $\bar{R}_3$ at which the mass is focused, then (A.8) and (A.10) give $R_3 \sim \bar{R}_3(1-\alpha^2/4)$. Also from (6.16), we have $R_2/R_3 \sim \alpha$, and from (6.21), we get $R'_3/R_3 =$

$m/m'$. So, from Figure 6.9, we can approximate

$$R'_{2f} = \left( (x + R_2 \cos \gamma)^2 + \left( \bar{R}_3 + R_2 \sin \gamma - R'_3 \right)^2 \right)^{1/2}$$

$$\sim \bar{R}_3 \left( 1 + \frac{x^2}{\bar{R}_3^2} \right)^{1/2} \tag{6.29}$$

Let us return to (6.27). From (6.11), (6.12), and (6.15), for $Z = 1$, we have $\bar{p}_3 \bar{R}_3 = m\alpha c \bar{R}_3 = \hbar$. Using this and (6.29) in (6.27), we arrive at

$$E'_{3f} \sim - \frac{Z' Z m \alpha^2 c^2}{\left( 1 + \frac{x^2}{\bar{R}_3^2} \right)^{\frac{1}{2}}} \tag{6.30}$$

Here, $m$ is the electron mass in motion.

Note that from (6.22) and (6.23), the magnitude of the spin-3 field-momentum $p'_{3f}$ is $\hbar \alpha / (R'_{2f}(1 - \alpha^2)^{1/2})$ where $\hbar = m\alpha c \bar{R}_3$ and the conversion to field energy in (6.30) requires its multiplication by $c(1 - \alpha^2)^{1/2}$, the spin-2 speed.

### 6.4.2.2 *Effect of electron field on proton ring*

We shall now see that the result (6.30) can be reproduced for the spin-3 effect of the electron ring on the proton target. The proton variables are primed. Equation (6.22) becomes

$$p_{2f} = \frac{\hbar}{R_{2f}} \tag{6.31}$$

where $R_{2f} = R'_{2f}$ in (6.29); see Figure 6.9. We have (6.23) for $Z$ electrons in the field source as

$$\vec{p}_{3f} = Z p_{2f} \frac{\alpha}{(1 - \alpha^2)^{1/2}} \begin{pmatrix} 0 \\ -\cos \theta \\ \sin \theta \end{pmatrix} \tag{6.32}$$

and (6.24) becomes

$$d\vec{R}'_2 = R'_2 d\gamma \begin{pmatrix} 0 \\ \cos \theta \\ -\sin \theta \end{pmatrix} \tag{6.33}$$

We now project the electron field momentum onto the proton ring to get

$$J_{3f} = Z' \int_{\gamma=0}^{\gamma=2\pi} \vec{p}_{3f} \cdot d\vec{R}_2' \qquad (6.34)$$

After dividing by the $Z'$-proton spin-2 time period $T_2'$ and noting that from (6.1), we have $2\pi R_2'/T_2' = c(1-\alpha^2)^{1/2}$, then after use of $\hbar = m\alpha c \bar{R}_3$, the approximation $R_3 \sim \bar{R}_3(1-\alpha^2/4)$, and (6.29), we find the electron field energy occupying the proton ring as

$$E_{3f} = -\frac{Z'Z\hbar\alpha c}{R_{2f}} \sim -\frac{Z'Zm\alpha^2 c^2}{\left(1+\frac{x^2}{R_3^2}\right)^{1/2}} \qquad (6.35)$$

Comparing (6.30) and (6.35), we have $E_{3f} = E_{3f}'$.

### 6.4.2.3  *Off-axis field effect*

Our treatment so far has focused on two interacting rings whose axes are coincident. We take this to be the condition for the electric field of one to affect the other. The result is a redistribution of ring action into linear action as motion along the $x$ axis. We now arrange for the axis of the source field to penetrate the space outside the aperture of the target ring; see Figure 6.10. In Figure 6.10(b), we can see clearly that by extending the source ring axis (dotted line) horizontally from A, it fails to pass through the target ring aperture at B. From (6.9) and (6.16), since the momentum field $\vec{p}_{3f} = \hat{\theta}k/r$, where $k$ is constant, then we must have $\nabla \times \vec{p}_{3f} = 0$, and by considering Stokes' theorem of vector calculus, we find that the source ring field has no effect on the target ring.

This affords us an insight into how the electric field of a source ring affects its target. Here, the electric field vector $\vec{E}$ can be considered to act along the OAM ring axis at $A$. Unless the axis of the source — and it might be curvilinear due to the presence of other fields — penetrates the aperture of a target ring, it has no effect.

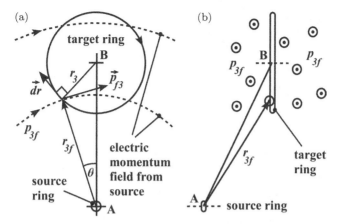

Figure 6.10 The effect of a source ring A momentum field on a target ring B, in which the extended source axis does not penetrate the target aperture. (a) The electric momentum-field $\vec{p}_{3f}$ passes through the target ring with inverse proportionality to the radius. The angle $\theta$ is taken in the plane of the page between the components of $r_{3f}$ and the line joining the ring centers AB. (b) Right-elevation view of (a) showing that the source axis at A does not penetrate the target aperture. The circled dots indicate electric field-momentum $p_{3f}$ passing out of the page.

## 6.5 Kinematics and Dynamics

### 6.5.1 *Law of momentum conservation*

It will now be shown how Newton's third law 'To every action there is always opposed an equal reaction' arises from MVR theory. Let us in time $\Delta t$ move the electron OAM ring a distance $\Delta x$ along the $x$ axis; see Figure 6.9(a). The change in spin-2 radius $r_{2f} = r'_{2f}$ will change the proton momentum field inhabiting the electron ring, but will also affect the electron momentum field intruding on the proton ring. If we consider the first case, from (6.30), we have

$$\frac{\Delta E'_{3f}}{\Delta x} \sim \frac{xZ'Zm\alpha^2c^2}{R_3^2\left(1 + \frac{x^2}{R_3^2}\right)^{\frac{3}{2}}} = \frac{\Delta p'_{3f}}{\Delta t} \qquad (6.36)$$

which we recognize as a force. We postulate that the electron ring tends to reduce the intruding action or energy. So, for an attractive force, we must have the change in proton–electron displacement

$\Delta x < 0$ for $\Delta E'_{3f}$ to decrease, so that the electron ring moves toward the proton. We note that $\Delta x / \Delta t$ is the speed with which the electron has been moved.

Let us now consider the electron momentum field in the proton ring; so we have from (6.35),

$$\frac{\Delta E_{3f}}{\Delta x} \sim \frac{x Z' Z m \alpha^2 c^2}{\bar{R}_3^2 \left(1 + \frac{x^2}{\bar{R}_3^2}\right)^{\frac{3}{2}}} = \frac{\Delta p_{3f}}{\Delta t} \tag{6.37}$$

Again, for oppositely rotating rings, we must have $\Delta x < 0$ to reduce the electron field energy in the proton ring, so that the rings move together in an attraction. For same-sense spin-3 rotations, the signs of (6.30) and (6.35) are opposite and we need $\Delta x > 0$ to effect an energy decrease. This separation amounts to a repulsion. Considering (6.36) and (6.37), in unit time, we have the change in electron field momentum in the proton ring equal to the change in proton field momentum in the electron ring; thus,

$$\Delta p_{3f} = \Delta p'_{3f} \tag{6.38}$$

Still in the center of mass frame, we now propose that an increase of field momentum in a ring, as should occur when the electron is moved toward the proton, displaces an equal spin-3 momentum out of both rings into linear momentum along the $x$ axis. This means that

$$\Delta p'_x = \Delta p_x \tag{6.39}$$

If we begin the coaxial proton and electron at an arbitrarily large distance $x \to \infty$ from the proton at rest, and give the electron an arbitrarily small velocity, we should find the sums of $\Delta p'_x$ and $\Delta p_x$ yielding

$$p'_x = p_x \tag{6.40}$$

where $p'_x$ and $p_x$ are the linear momenta for the single proton and electron, respectively. This is the law of momentum conservation for two mass vortex rings electrically interacting in an unbound state.

### 6.5.2 The reduced mass

The derivation of the reduced mass is a well-known argument based on the equality of the proton and electron linear momentum relative to the center of mass. Using (6.31) and (6.32), the magnitude of the single-electron spin-3 field momentum inhabiting the single-proton ring is obtained from $Z' = Z = 1$ so that

$$p_{3f} = p'_{3f} = \frac{\hbar\alpha}{R_{2f}(1 - \alpha^2)^{1/2}} \tag{6.41}$$

where $R_{2f} = R'_{2f}$ is given by (6.29). However, as stated earlier, the intruding field momentum displaces an equivalent resident ring momentum into linear momentum, and (6.41) demonstrates that the magnitudes of the two displaced momenta are the same but with opposite signs; so using (6.40), we can establish

$$m\vec{v} + m'\vec{v}' = 0 \tag{6.42}$$

Our aim is to express the following total kinetic energy $K$ in terms of $\dot{R}$ and $\dot{r}$, the non-relativistic speeds of the center of mass and the electron relative to the proton, respectively:

$$K = \frac{1}{2}m\vec{v} \cdot \vec{v} + \frac{1}{2}m'\vec{v}' \cdot \vec{v}' \tag{6.43}$$

where $\vec{v}' = d\vec{r}_1/dt$ and $\vec{v} = d\vec{r}_2/dt$ (for this, we temporarily abandon our previous definitions of $R$ and $r$). Using (6.42), our two premises are

$$m\left(\frac{d\vec{R}}{dt} - \frac{d\vec{r}_2}{dt}\right) + m'\left(\frac{d\vec{R}}{dt} - \frac{d\vec{r}_1}{dt}\right) = 0 \tag{6.44}$$

and the geometrical relation, see Figure 6.11,

$$\frac{d\vec{r}}{dt} = \frac{d\vec{r}_2}{dt} - \frac{d\vec{r}_1}{dt} \tag{6.45}$$

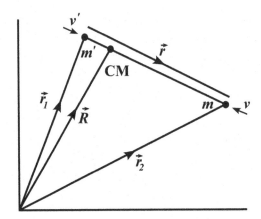

Figure 6.11    A two-body system consisting of a proton mass $m'$ and an electron mass $m$ with a center of mass (CM). They are traveling toward the center of mass at speeds $v'$ and $v$, respectively.

The following result can be shown to be equivalent to (6.43) by substituting $d\vec{R}/dt$ into it from (6.44) and $d\vec{r}/dt$ from (6.45):

$$K = \frac{1}{2}\left(m' + m\right)\frac{d\vec{R}}{dt}\cdot\frac{d\vec{R}}{dt} + \frac{1}{2}\frac{m}{\left(1 + \frac{m}{m'}\right)}\frac{d\vec{r}}{dt}\cdot\frac{d\vec{r}}{dt} \qquad (6.46)$$

If the velocity of the center of mass $d\vec{R}/dt = 0$, then all that remains is the second term on the right side which is the kinetic energy of the electron with respect to the proton with the reduced mass

$$\mu = \frac{m}{\left(1 + \frac{m}{m'}\right)} \qquad (6.47)$$

For a general mass, we need only make the substitutions $m \to Zm$ and $m' \to Z'm'$.

### 6.5.3    *Ring energy invariance*

We have already discussed the invariance of the total energy in a target ring for two coaxially interacting rings. This is the sum of the external field energy and the remaining ring energy in the target ring. There is a need to quantify the two types of field interaction with the

OAM mass ring: repulsion and attraction. In both cases, the electron ring-plane's spin-3 field-momentum $\Delta p'_{3f}$ that arises from the proton adds to the electron's resident ring momentum $p_3$ and overloads the ring. Let us increase time $\Delta t > 0$ and adapt (6.36) for a repulsion by inserting a negative sign to produce (6.48).[22] In order to reduce the field energy $\Delta E'_{3f} < 0$ (or field-momentum $\Delta p'_{3f} < 0$) in the ring, we need the right side to be negative, which occurs when $\Delta x > 0$, that is, the source and target rings move apart; see Figure 6.9. An unrestrained ring should tend to redirect this additional field momentum into linear momentum along the $x$ axis; thus,

$$\frac{\Delta E'_{3f}}{\Delta x} = \frac{\Delta p'_{3f}}{\Delta t} = -\frac{xZ'Zm\alpha^2c^2}{R_3^2\left(1 + \frac{x^2}{R_3^2}\right)^{\frac{3}{2}}} \tag{6.48}$$

For an attractive force, the negative sign on the right side of (6.48) is replaced by a positive one, and this time the reduction in field energy or momentum is obtained from $\Delta x < 0$. The source and target rings move toward each other. So, an amount of electron momentum, equal to the field momentum in the target ring plane, is displaced into linear momentum along the $x$ axis. This brings out the essence of the field which does not change the energy of the OAM ring, but controls the partition of its constant value between resident ring-plane energy and intruding field energy.

### 6.5.4 *Active and passive acceleration*

The present author first suggested that acceleration should have two forms — active and passive — in *The Quantum Puzzle* (Clarke, 2017, 241–247).[23] The following considerations should illustrate why a single type of acceleration does not cover all possibilities. Allow there to be an OAM mass ring $S'$ at rest in our observation frame $S$ which has an identical mass ring also at rest; see Figure 6.12(a).

---

[22]This means the same-sense rotation for (6.23) and (6.24) and the vectors must be the same.

[23]However, the analysis focused incorrectly on an SAM ring and a determining speed $c$.

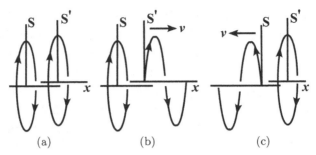

Figure 6.12    Active and passive acceleration. (a) Frames $S$ and $S'$ at rest with respect to each other. (b) In $S$, frame $S'$ absorbs energy and is actively accelerated with speed $v$ along the positive x axis. (c) In $S'$, frame $S$ does not absorb energy and is passively accelerated with speed $v$ along the negative $x$ axis.

Each has a normal vector along the $x$ axis. In $S$, let the ring $S'$ absorb energy so that it accelerates to a speed $v$ along the positive $x$ axis; see Figure 6.12(b). The SAM cluster now takes on a helical trajectory. In consequence, we expect that the mass of the ring in $S'$ should increase. This type of acceleration is well known but shall be denoted 'active acceleration' here in virtue of the fact that the absorption of energy has produced a physical change in the OAM ring $S'$.[24]

Let us now observe the ring $S$ from the physically changed $S'$. No energy has been observed entering $S$, yet it accelerates along the negative $x$ axis with the same speed $v$; see Figure 6.12(c). Let us denote this new type of acceleration 'passive acceleration' to reflect the circumstance that it accelerates without an exchange of energy with its environment. This is a case that is given little attention in modern dynamics. Without absorption or emission, there are no grounds for claiming that the mass of the ring varies.

What does one frame observe the mass of the other to be? If the OAM ring in the actively accelerated $S'$ frame were to explode and $S$ were to collect and measure all the radiation energy, there is no doubt that $S$ must expect from the knowledge that $S'$ had absorbed energy that it had increased along with its mass. What if

---

[24]The suggestion here is that the entire OAM ring moves rather than having changes in its internal energy levels.

$S'$ were to collect and measure all the energy from the explosion of the passively accelerated $S$ ring? The question would be straightforward to answer had the measuring frame $S'$ not absorbed energy for then it would not have physically changed. The complication is that $S'$ *has* absorbed energy and that must have resulted in a physical change in its measuring instruments. For example, with an increase in the SAM cluster mass in the $S'$ ring which determines internal energy level transitions (see Chapter 7), an excitation from a lower to a higher energy level must now require more energy than before, even for a nuclear excitation. So, if the explosion of $S$ when the two frames were at rest with respect to each other provided the exact amount of energy $\varepsilon$ needed to produce a particular nuclear transition in an atom in $S'$, when $S'$ is set into motion in an active acceleration, this energy $\varepsilon$ from $S$ would now be too small. So, $S'$ must conclude that the passively accelerated $S$ has reduced in energy and mass, despite its lack of energy exchange with its surroundings.

We could also connect the frequency of the radiation energy from the explosion with time intervals. The frame $S'$ measures the emitted frequency from $S$ to decrease and so its time interval has increased and time has slowed. The frame $S$ observes the emitted frequency from $S'$ to have increased and so the time interval has decreased and time has speeded up. Surely, the resolution of the twin paradox lies in this fascinating comparison of passively and actively accelerated reference systems. The essence of it is that we can no longer maintain the concept of a 'rigid rod'.

The passively accelerated frame $S$ could just have easily experienced an attractive or repulsive field that would not have changed its energy. So let us now consider the kinematics of the active and passive accelerations. First let us define

$$\Lambda = \left(1 - \left(\frac{v_x}{\alpha c}\right)^2\right)^{1/2} \tag{6.49}$$

Figure 6.13 is a representation of Figures 6.12(b)(c). We can see an increment of spin-3 momentum on the target ring for the SAM cluster, represented by the vertical line of the triangle in the ring plane. The horizontal line is parallel to the OAM ring normal and

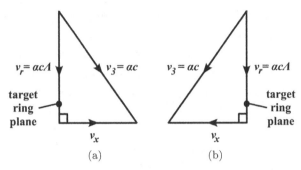

Figure 6.13    The kinematics of an 'active' and 'passive' acceleration examined using an increment of spin-3 momentum on an OAM mass ring. The hypotenuse is an increment of the path that the SAM cluster takes through the reference frame. (a) Active acceleration for $S'$, (b) passive acceleration for $S$.

carries the OAM mass ring linear speed $v_x$ along the $x$ axis. In both cases, the SAM cluster in the spin-3 target ring takes a longer path through the stationary center of mass frame, retaining speed $v_3 = \alpha c$. This is represented by the hypotenuse of the triangle. The active and passive cases are identical in their kinematics, mirror the cases of repulsive and attractive electrical interaction, and only differ in the directions that the rings $S'$ and $S$ move.

However, their dynamics are not identical. Let $m$ be the mass of the HSD cluster rotating round the stationary field-free OAM mass ring. For Figure 6.14(a), the 'active' acceleration in which energy is absorbed results in an increased mass by multiplying all sides of the triangle in Figure 6.13(a) by $m/\Lambda$. This retains the momentum $m\alpha c$ in the plane of the ring, but the absorption of radiation has resulted in an increased momentum along the helical trajectory of $m\alpha c/\Lambda$. In Figure 6.14(b), there is no absorption of energy and no increase of momentum represented in the helical trajectory, so we multiply all sides of the triangle in Figure 6.13(b) by $m$. All that has happened here is that the momentum in the plane of the ring has been diverted into a helical path.

In consequence, we can state the invariance law for 'active' and 'passive' acceleration as follows:

*Invariance law for active acceleration*

$$(m\alpha c)^2 = (m_r \alpha c)^2 - (m_r v_x)^2 \tag{6.50}$$

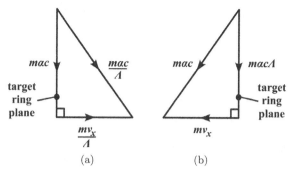

Figure 6.14  The dynamics of an 'active' and 'passive' acceleration examined with an increment of spin-3 momentum on an OAM mass ring. The hypotenuse is the path that the SAM cluster takes through the reference frame. (a) Active acceleration for $S'$, (b) passive acceleration for $S$.

*Invariance law for passive acceleration*

$$(m\alpha c)^2 = (m\alpha c\Lambda)^2 + (mv_x)^2 \tag{6.51}$$

where in Equation (6.50), the mass in the ring plane is

$$m_r = \frac{m}{\Lambda} = \frac{m}{\left(1 - \left(\frac{v_x}{\alpha c}\right)^2\right)^{1/2}} \tag{6.52}$$

According to the 'active' case, in Equation (6.52) the limiting speed for an OAM mass ring is $v_x = \alpha c$.

From the 'passive' case (6.51), if we set the mass in the ring plane as $m_r = m\Lambda$, then

$$m_r = m\left(1 - \left(\frac{v_x}{\alpha c}\right)^2\right)^{1/2} \tag{6.53}$$

The 'passive' acceleration case imitates the relation of a ring to an external electric field, both for attraction and repulsion, in that there is no energy exchange with the field, only a diversion of momentum from the ring plane into linear motion along its axis, and the field-free stationary OAM contains the maximum amount of momentum in the ring plane that can be diverted away from it. Equation (6.53) suggests that there is a limiting speed $v_x = \alpha c$ with which an electron ring can approach a proton ring in an electrical attraction; see Chapter 8. In contrast, 'active' acceleration requires an absorption or emission of radiation by the ring, a physical change in its structure that results in a change in mass. We shall now suggest a mechanism for active acceleration absorption for OAM ring motion and compare that with a mechanism for absorption that produces internal energy level transitions.

### 6.5.5   Ring–radiation interaction

We now address the question as to how radiation interacts with the OAM mass ring. The proposal shall be that if the rotation sense of the radiation is the same as the ring, then it overloads the ring with energy and redistributes it into linear momentum. The ring then recedes from the radiation source, see Figure 6.15(a), and the OAM mass ring increases in mass and acquires external energy in an 'active' acceleration. On the other hand, if the radiation rotation opposes that of the ring, then there is no effect on the linear motion of the ring. Instead, the SAM cluster takes on a higher-radius energy level as the energy of the SAM rotation increases; see Figure 6.15(b).[25] We then have a 'loaded' OAM mass ring. This shall be the focus of Chapter 7.

### 6.5.6   Representations of interactions

There is a way of representing the momentum relations of the following: (a) radiation absorption for external ring motion, and (b) radiation absorption for internal energy transitions.

While the energy $\varepsilon$ of a light ray is obtained from its momentum $p$ by a multiplication by $c$ as $\varepsilon = pc$, the energy in an OAM mass ring at rest is obtained from its momentum by a multiplication by $\alpha c$, the spin-3 ring-plane speed, as $\varepsilon = p\alpha c$. If the OAM ring is in motion, we can expect the multiplier $\alpha c$ to diminish relativistically.

#### 6.5.6.1   Radiation absorption for external ring motion

In Figure 6.16(a), the radiation incident on the OAM ring has the same rotation sense, so it results in linear motion of the ring along the

---

[25]The assumption is that as the radius increases, the energy decreases in magnitude but increases by becoming less negative.

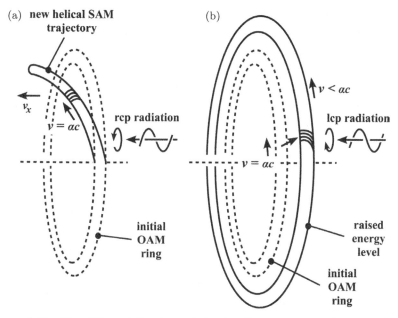

Figure 6.15   The effect of circular polarized radiation on a clockwise rotating OAM mass ring as seen from the right. (a) The right-circular polarized radiation and the ring share the same rotation sense, so the added energy is redistributed into linear motion of the ring, that is, external energy. (b) The left-circular polarized radiation and the ring have opposite rotation so the SAM cluster takes on energy internally and is raised to a higher-radius energy level.

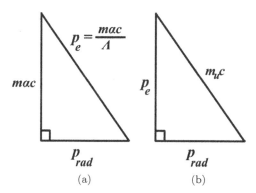

Figure 6.16   Relationship of the resulting ring-plane momentum with: (a) radiation momentum that causes OAM ring motion, (b) radiation momentum that causes internal level transitions.

direction of its axis with momentum $p_e$. The field-free stationary ring momentum $m\alpha c$ is retained in the ring plane and the radiation momentum is set orthogonal to it to produce an increased total momentum $m\alpha c/\Lambda$. This mirrors the 'active' acceleration case depicted in Figure 6.14(a). Here, the connection is $p_{\text{rad}} = mv_x/\Lambda$, that is, the radiation momentum is entirely displaced into an increased mass linear momentum of the ring. As suggested in Section 6.5.5, the rotation sense of the incident radiation would need to be the same as that of the OAM ring for this to occur.[26] It is further suggested that an electron in an excited state can be sent to a lower radius state by stimulated emission when the radiation and OAM ring rotation senses are identical.

### 6.5.6.2 *Radiation absorption for internal energy transitions*

This is where the incident radiation and the OAM ring have opposite rotation sense; see Figure 6.16(b). It has similarities with Figure 6.16(a). Again the remaining SAM momentum $p_e$ in the ring plane is set orthogonal to the external momentum $p_{\text{rad}}$. However, whereas in Figure 6.16(a) the field momentum is diverted into linear momentum of the entire ring along the direction of its axis, here the radiation momentum is diverted internally. Also, with the emphasis purely on radiation interaction (i.e., between external radiation and the SAM mass-radiation structure) rather than the linear motion of the OAM ring, the resultant momentum is now $mc$ instead of $m\alpha c$. This process is represented as (7.1) in Chapter 7 for the fine-structure calculation.

In conclusion, an external field does not exchange energy with an OAM ring, it only redirects it. Both an attractive and a repulsive interaction displace momentum out of the ring plane into linear motion of the ring. Incident radiation acting on an OAM ring in

---

[26]In fact, on the basis of the same-sense rotation of radiation and ring for the transmission of external ring momentum, I originally entertained the notion that a repulsion field, which has the same-sense rotation as the OAM ring, was an example of 'active' acceleration. However, I could not convince myself that the field exchanged energy with the ring.

ground state will create external or linear motion of the ring if the radiation and OAM ring rotation senses are the same. If the senses are opposite, the SAM mass ring will be raised to a higher radius, that is, there will be an internal energy change. An OAM ring in a raised internal energy state can be reduced to a lower state by a same-sense rotation in the radiation absorption. This is a stimulated emission.

## References

Allen, L., M. W. Beijersbergen, R. J. C. Spreeuw, and J. P. Woerdman. 'Orbital angular momentum of light and the transformation of Laguerre–Gaussian laser modes'. *Physical Review A*, 45 (1992): 8185–8189.

Beijersbergen, M. W., R. P. C. Coerwinkel, M. Kristensen, and J. P. Woerdman. 'Helical wavefront laser beams produced with a spiral phaseplate'. *Optics Communications*, 112 (1994): 321–327.

Clarke, B. R. *The Quantum Puzzle: A Critique of Quantum Theory and Electrodynamics*. Singapore: World Scientific Publishing, 2017.

Friese, M. E. J., J. Enger, H. Rubinsztein-Dunlop, and N. R. Heckenberg. 'Optical angular-momentum transfer to trapped absorbing particles'. *Physical Review A*, 54 (1996): 1593–1596.

Jammer, M. *Concepts of Mass in Classical and Modern Physics*. New York: Dover Publications, 1997.

Leach, J., S. Keen, M. J. Padgett, C. Saunter, and G. D. Love. 'Direct measurement of the skew angle of the Poynting vector in a helically phased beam'. *Optics Express*, 14 (2006): 11919–11924.

McMorran, B. J., A. Agrawal, I. M. Anderson, A. A. Herzing, H. J. Lezec, J. J. McClelland, and J. Unguris. 'Electron vortex beams with high quanta of orbital angular momentum'. *Science*, 331 (2011): 192–195.

Padgett, M., and L. Allen. 'Light with a twist in its tail'. *Contemporary Physics*, 41 (2000): 275–285.

Padgett, M., J. Courtial, and L. Allen. 'Light's orbital angular momentum'. *Physics Today*, May (2004): 35–40.

Uchida, M., and A. Tonomura. 'Generation of electron beams carrying orbital angular momentum'. *Nature*, 464 (2010): 737–739.

Verbeeck, J., H. Tian, and P. Schattschneider. 'Production and application of electron vortex beams'. *Nature*, 467 (2010): 301–304.

## Chapter 7

# The Loaded OAM Mass Ring

*The orbital angular momentum (OAM) mass ring is extended to include rings that have absorbed radiation which we denote as 'loaded' rings. Experimental evidence shows that such rings can have energy levels without the influence of an external potential. A new derivation of the Sommerfeld–Dirac fine-structure formula is obtained using a self-potential, and the properties of elliptical rings are examined. After an investigation into the structure of the momentum field, a derivation of Coulomb's law is given for excited states.*

## 7.1 Overview

In Section 6.1, the notion of a 'loaded' mass ring was introduced, where an orbital angular momentum (OAM) ring can have energy levels in a field-free region, that is, in the absence of an external electric potential. Radiation can gain access to its energy levels, a circumstance that occurs when the rotation sense of its circular polarization or optical spin angular momentum is in opposition to that of the OAM ring; see Section 6.5.6.2. It has also been suggested that stimulated emission might occur when an excited ring is targeted with radiation in the same direction of rotation.

An external attractive electric potential permeating a target ring displaces the equivalent target-ring energy into linear motion, and determines the proportion that remains in the ring to participate in energy transitions; see Section 6.5.6.1. In this sense, the position of the ring in the electric potential field affects the magnitude of the energy transitions. We posit here that the ring emits energy in an electric potential field when the field energy intruding in the electron ring plane and the remaining SAM energy in the ring are equal, that is, each has energy $ma^2c^2/2$. This occurs at the oscillation boundary and will be the basis of the bound state to be developed in which Doppler-shift (Chapter 8) and hyperfine corrections (Chapter 9) will also be applied.

Our present aim is to set up the basis for the bound-state treatment in Chapter 8. Without recourse to an external potential, an alternative derivation of the Sommerfeld–Dirac fine-structure formula for the hydrogen atom spectrum is produced using the above-mentioned principle of equal proton-field and resident electron energies in the electron ring. We shall also examine the field energy from a source ring that inhabits a target ring. The loaded ring radius will now be given in lower case, for example, spin-3 radius $r_3$, in contrast to the upper case that has previously been used for the unloaded ring $R_3$. The properties of elliptic OAM rings will also be explored.

## 7.2   Sommerfeld–Dirac Fine Structure

### 7.2.1   *Mass dependence on quantum numbers*

We now introduce a modified Sommerfeld-type calculation for the fine structure energy levels of the hydrogen atom. We posit that the effect of absorbing radiation with momentum $p_{\text{rad}}$ in an induced absorption will be to reduce to mass $m$ the average spin-3 mass on the left of (6.18), which we denote by $m_u$ as follows:

$$(mc)^2 = (m_u c)^2 - (p_{\text{rad}})^2 \tag{7.1}$$

This is represented in Figure 6.16(b) where $p_e = mc$, and $m$ is the resulting mass. If $p_{\text{rad}}$ includes both an azimuthal momentum $n_\phi \hbar / r_{3f}$, $n_\phi \in \mathbb{Z}^+$, and a radial momentum $p_r$ (whose directions are mutually orthogonal), then using (6.18) with a variable spin-3 field radius $r_{3f}$ instead of $\bar{R}_3$ confined to the ring, we have the following momentum relation:

$$(mc)^2 = \left( m_o c + \frac{\hbar \alpha}{r_{3f}} \right)^2 - \left( \frac{n_\phi \hbar}{r_{3f}} \right)^2 - (p_r)^2 \tag{7.2}$$

For an unloaded OAM mass ring, the last two terms of (7.2) vanish and the momentum of the SAM mass is again $mc = m_u c$, as given by (6.18). We also recall that $m_o = m_{2o}$ is defined by the spin-2 radius in (6.3). We make the further assumption that this rest mass $m_o$ does not vary over the energy levels that the OAM mass ring adopts; see (7.26).

Rearranging (7.2) and confining the self-potential radius $r_{3f}$ to the OAM ring as $r_3$ yields

$$p_r = \left( \frac{\hbar^2(\alpha^2 - n_\phi^2)}{r_3^2} + \frac{2m_o\hbar\alpha c}{r_3} + (m_o^2 - m^2)c^2 \right)^{1/2} \tag{7.3}$$

We now set up the action integral with the radial quantum number $n_r \in \mathbb{Z}^+$, as follows:

$$\oint p_r dr_3 = n_r h \tag{7.4}$$

Sommerfeld (1923, 551–552) has already given the following mathematical result:

$$\oint \left( \frac{C}{r_3^2} + \frac{2B}{r_3} + A \right)^{1/2} dr_3 = -2\pi i \left( \sqrt{C} - \frac{B}{\sqrt{A}} \right) = n_r h \tag{7.5}$$

In the present case, after comparison of (7.3) and (7.4) with (7.5), and noting that $i(\alpha^2 - n_l^2)^{1/2} = (n_\phi^2 - \alpha^2)^{1/2}$ and $i/(m_o^2 - m^2)^{1/2} = -1/(m^2 - m_o^2)^{1/2}$, we find

$$m = \frac{m_o}{\left( 1 - \frac{\alpha^2}{Y^2\left(1+\frac{\alpha^2}{Y^2}\right)} \right)^{1/2}} = m_o \left( 1 + \frac{\alpha^2}{Y^2} \right)^{1/2} \tag{7.6}$$

where $Y = n_r + (n_\phi^2 - \alpha^2)^{1/2}$.[1] This differs from (7.7), which is Sommerfeld's fine-structure equation (24), in which the sign of the

---

[1] Equation (7.6) at $n_r = 0$ and $n_\phi = 1$ appears to give the correct mass in Principle 7.15 in Clarke (2017, 254). However, this principle is intended for an unloaded ring and so is in error.

half power is altered (Sommerfeld 1923, 473); see Appendix B[2]:

$$m = m_o \left(1 + \frac{\alpha^2}{Y^2}\right)^{-1/2} \tag{7.7}$$

In consequence, one might imagine that (7.6) is in error; however, we shall see that it is precisely the required form to set up the new MVR hydrogen atom calculation.

### 7.2.2   *Spin-2 and spin-3 speed*

We note that for $n_r = 0$ and $n_\phi = 1$, (7.6) becomes

$$m = \frac{m_o}{(1 - \alpha^2)^{\frac{1}{2}}} \tag{7.8}$$

A comparison of (7.6) and (7.8) suggests that the azimuthal speed $v_3$ transforms from the ground state $(n_r, n_\phi) = (0, 1)$ to general quantum numbers $(n_r, n_\phi)$ as follows:

$$v_3 = \alpha c \rightarrow \frac{\alpha c}{Y \left(1 + \frac{\alpha^2}{Y^2}\right)^{1/2}} \tag{7.9}$$

Since we shall demand for all OAM mass rings that $v_2^2 + v_3^2 = c^2$, then the corresponding spin-2 speed must be

$$v_2 = \frac{c}{\left(1 + \frac{\alpha^2}{Y^2}\right)^{1/2}} \tag{7.10}$$

For motion on an elliptic path, $n_r > 0$, there is no radial speed at the perigee and apogee, and for ellipses the azimuthal speed at the former is greater than that at the latter. This suggests a variation in $v_3$. However, (7.9) is a constant of the motion on an OAM ring. We need to remind ourselves that in addition to energy of motion, there is a self-potential energy — see (6.18) — connected with the spin-3 radius and it is suggested that this acts in such a way as to

---

[2]Sommerfeld has $n'$ and $n_\phi$ for his radial and azimuthal quantum numbers, respectively. Here, we use $n_r$ and $n_\phi$. Appendix B gives Sommerfeld's mass, and azimuthal radius and action results.

maintain speed constancy. The effect of the curvature on $v_3$ will not be investigated further here but it appears that a reduction in spin-3 radius $r_3$ also reduces $v_3$, see Table 9.1. For this reason, we should expect the fine-structure constant $\alpha$ to be represented by a modified constant $\alpha' = Kalpha \times \alpha$, where $Kalpha < 1$.

### 7.2.3   *Spin-2 and spin-3 momentum*

From (7.6) and (7.10), the spin-2 momentum can be shown to be independent of $(n_r, n_\phi)$ as follows:

$$p_2 = mv_2 = m_o \left(1 + \frac{\alpha^2}{Y^2}\right)^{1/2} \frac{c}{\left(1 + \frac{\alpha^2}{Y^2}\right)^{1/2}} = m_o c \qquad (7.11)$$

where $m = m_2$ can be used since the average radius $\bar{r}_3$ has been chosen in such a way that the average spin-3 kinetic mass equals the spin-2 kinetic mass.[3] For the average spin-3 momentum at the average radius $\bar{r}_3$,[4] we have

$$\bar{p}_3 = mv_3 = m_o \left(1 + \frac{\alpha^2}{Y^2}\right)^{1/2} \frac{\alpha c}{Y \left(1 + \frac{\alpha^2}{Y^2}\right)^{1/2}} = \frac{m_o \alpha c}{Y} \qquad (7.12)$$

As a consequence of (7.11) and (7.12), we find that

$$\frac{\bar{p}_3}{p_2} = \frac{\alpha}{Y} \qquad (7.13)$$

---

[3]Principle 7.5 in Clarke (2017, 236) is in error, in that the spin-1 sense does not affect the linear momentum that participates in the OAM ring trajectories. For spin-2 field momentum, the notion in Principle 7.6 that "its curvature [around the optic axis] generates [causes] a vortex field" has now been superseded by the idea that it is a preexisting array of helical space dislocations (HSDs) that rotates around the spin-2 axis, the stretching out of which constitutes the field.

[4]The spin-3 momentum is the product of mass and speed. The spin-3 speed does not vary over the spin-2 circuit but due to angular momentum conservation the self-potential mass changes as $r_{3f}$ varies. So, taken over the spin-2 circuit, we demand that the average radius $\bar{r}_3$ be chosen so that the average spin-3 mass is the same as the spin-2 mass, that is, $\bar{m}_3 = m_2 = m$. The use of an average mass necessitates the use of an average momentum $\bar{p}_3$.

and adding $p_2^2 + \bar{p}_3^2$ (taken on the ring) then dividing by $c^2$ recovers (7.6).[5]

### 7.2.4    *Energy of SAM ring motion in an OAM mass ring*

Using (7.6) and (7.9), the energy contained in the spin-3 component on a stationary OAM mass ring in a field-free region is posited to be

$$\mathcal{E} = \bar{p}_3 v_3 = mv_3^2 = m_o \left(1 + \frac{\alpha^2}{Y^2}\right)^{1/2} \frac{\alpha^2 c^2}{Y^2 \left(1 + \frac{\alpha^2}{Y^2}\right)}$$

$$= m_o \frac{\alpha^2 c^2}{Y^2 \left(1 + \frac{\alpha^2}{Y^2}\right)^{1/2}} \tag{7.14}$$

where $Y = n_r + (n_\phi^2 - \alpha^2)^{1/2}$. The justification for this choice is that from (7.6), (7.9), and (7.10), we can obtain the total spin-1 energy, $mc^2 = mv_2^2 + mv_3^2$. This suggests that the spin-2 and spin-3 energies are the two additive mutually orthogonal components of the resultant spin-1 light ray energy.

### 7.2.5    *Sommerfeld–Dirac energy levels*

Using (7.6), let us now rewrite (7.14) as

$$\mathcal{E} = \frac{m_o}{\left(1 - \frac{v_3^2}{c^2}\right)^{1/2}} v_3^2 = \frac{m_o}{\left(1 - \frac{\alpha^2}{Y^2\left(1 + \frac{\alpha^2}{Y^2}\right)}\right)^{1/2}} \frac{\alpha^2 c^2}{Y^2 \left(1 + \frac{\alpha^2}{Y^2}\right)} \tag{7.15}$$

We shall now carry out a transformation on momentum (7.12), namely, $\bar{p}_3 \to \bar{p}_3/\sqrt{2}$, that will lead us to an alternative derivation of the fine-structure formula.[6] This will embrace our requirement — to be developed in Chapter 8 — that in the electron ring plane

---

[5]Principle 7.17 in Clarke (2017, 255) gives the correct ratio of $\bar{p}_3/p_2$ for $n_r = 0$ and $n_\phi = 1$ in (7.13) above.

[6]This momentum transformation representing emission at the oscillation boundary of the hydrogen atom was omitted when formulating (9.58) in Clarke (2017, 334).

the intruding proton field energy and the remaining electron spin-3 energy are equal at the point of radiative emission. So,

$$v_3 = \frac{\alpha c}{Y\left(1 + \frac{\alpha^2}{Y^2}\right)^{1/2}} \rightarrow \frac{\alpha c}{\sqrt{2}Y\left(1 + \frac{\alpha^2}{Y^2}\right)^{1/2}} \tag{7.16}$$

Using (7.16), this turns (7.15) into

$$\mathcal{E} = \frac{m_o}{\left(1 - \frac{\alpha^2}{2Y^2\left(1+\frac{\alpha^2}{Y^2}\right)}\right)^{1/2}} \frac{\alpha^2 c^2}{2Y^2\left(1 + \frac{\alpha^2}{Y^2}\right)} \tag{7.17}$$

For an expansion in powers of $\alpha$, we note that

$$Y^2 = \left(n_r + \left(n_\phi^2 - \alpha^2\right)^{1/2}\right)^2 = n^2\left(1 - \frac{\alpha^2}{nn_\phi} - \frac{\alpha^4 n_r}{4n^2 n_\phi^3} + \cdots\right) \tag{7.18}$$

where $n = n_r + n_\phi$, and consequently obtain the expansion of (7.17) as

$$\mathcal{E} = \frac{m_o \alpha^2 c^2}{2Y^2}\left(1 + \frac{\alpha^2}{4Y^2}\left(1 - \frac{\alpha^2}{Y^2}\right) + \frac{3\alpha^4}{32Y^4} + \cdots\right)$$
$$\times \left(1 - \frac{\alpha^2}{Y^2} + \frac{\alpha^4}{Y^4} + \cdots\right) \sim \frac{m_o \alpha^2 c^2}{2Y^2}\left(1 - \frac{3\alpha^2}{4Y^2} + \frac{19\alpha^4}{32Y^4} + \cdots\right) \tag{7.19}$$

Using (7.18), up to sixth order in $\alpha$, this becomes

$$\mathcal{E} = \frac{m_o \alpha^2 c^2}{2n^2}\left(1 + \frac{\alpha^2}{n^2}\left(\frac{n}{n_\phi} - \frac{3}{4}\right)\right.$$
$$\left. + \frac{\alpha^4}{n^4}\left(\frac{n_r n^2}{4n_\phi^3} + \frac{n^2}{n_\phi^2} - \frac{3n}{4n_\phi} + \frac{19}{32}\right) + \cdots\right) \tag{7.20}$$

which agrees with the Sommerfeld–Dirac fine-structure formula to fourth order in $\alpha$; see White (1934, 134). There is also good agreement to sixth order: of these four terms, in the first, White gives an extra $n$ in place of $n_r$; the coefficient of the second term as

Table 7.1    A selection of fine-structure energy levels (MHz) of atomic hydrogen as given by the Sommerfeld–Dirac theory, the new MVR theory, and by experiment; see Horbatsch and Hessels (2016, Tables 3–5) at *Kalpha* = 1.

| Level | $n$ | $L$ | $Sp$ | $n_\phi$ | $n_r$ | Sommerfeld | MVR | Expt *h-f* average |
|-------|-----|-----|------|----------|-------|------------|-----|--------------------|
| $1S_{1/2}$ | 1 | 0 | +1/2 | 1 | 0 | 3 288 095 006.03 | 3 288 094 981.92 | 3 288 087 212.21 |
| $2S_{1/2}$ | 2 | 0 | +1/2 | 1 | 1 | 822 026 487.44 | 822 026 485.94 | 822 025 488.31 |
| $2P_{1/2}$ | 2 | 1 | −1/2 | 1 | 1 | 822 026 487.44 | 822 026 485.94 | 822 026 516.55 |
| $2P_{3/2}$ | 2 | 1 | +1/2 | 2 | 0 | 822 015 543.74 | 822 015 542.25 | 822 015 535.67 |
| $3S_{1/2}$ | 3 | 0 | +1/2 | 1 | 2 | 365 343 889.55 | 365 343 889.26 | 365 343 591.60 |
| $3P_{1/2}$ | 3 | 1 | −1/2 | 1 | 2 | 365 343 889.55 | 365 343 889.26 | 365 343 897.71 |
| $3P_{3/2}$ | 3 | 1 | +1/2 | 2 | 1 | 365 340 646.99 | 365 340 646.68 | 365 340 644.11 |
| $3D_{3/2}$ | 3 | 2 | −1/2 | 2 | 1 | 365 340 646.99 | 365 340 646.68 | 365 340 649.09 |
| $3D_{5/2}$ | 3 | 2 | +1/2 | 3 | 0 | 365 339 566.16 | 365 339 565.85 | 365 339 565.45 |
| $4S_{1/2}$ | 4 | 0 | +1/2 | 1 | 3 | 205 505 424.89 | 205 505 424.79 | 205 505 298.86 |
| $4P_{1/2}$ | 4 | 1 | −1/2 | 1 | 3 | 205 505 424.89 | 205 505 424.79 | 205 505 428.23 |
| $4P_{3/2}$ | 4 | 1 | +1/2 | 2 | 2 | 205 504 056.94 | 205 504 056.83 | 205 504 055.62 |
| $4D_{3/2}$ | 4 | 2 | −1/2 | 2 | 2 | 205 504 056.94 | 205 504 056.83 | 205 504 057.76 |
| $5S_{1/2}$ | 5 | 0 | +1/2 | 1 | 4 | 131 523 239.92 | 131 523 239.88 | 131 523 175.31 |
| $6S_{1/2}$ | 6 | 0 | +1/2 | 1 | 5 | 91 335 465.75 | 91 335 465.72 | 91 335 428.31 |

*Note*: The MVR theory has a relativistic correction in the electron reduced mass. The experimental value is calculated from the mid-point of the two hyperfine values, and emerges from the BASIC program output as item (15); see Appendix D.

3/4 instead of 1; the coefficient of the third term as −3/2 instead of −3/4; and the last term as 5/8 instead of 19/32.

The reduced mass (6.47) is included in the Sommerfeld–Dirac calculation, while the MVR calculation uses it with a relativistic electron mass taken from (7.17). A selection of these levels is given in Table 7.1 taken from output items (1) and (2) using the BASIC computer program in Appendix C. These are compared with the midpoint of the hyperfine energies, item (15), calculated from Horbatsch and Hessels (2016, Tables 3–5).[7] All energies are given in MHz. The quantum variables which govern the new vortex calculation are $n_\phi = L + Sp + 1/2$ and $n_r = n - n_\phi$.

---

[7]The center of gravity of the fine-structure National Institute of Standards and Technology (NIST) values given by Kramida (2010, Table 4) are calculated from $E_{c.g.} = \sum_F (2F + 1)E_F / \sum_F (2F + 1)$ and are taken in relation to a zero ground state energy. This calculation is not used here.

### 7.2.6 *Elliptical OAM rings*

If we return to (7.3) and solve for the radial momentum $p_r = 0$, then the maximum and minimum spin-3 radii of the OAM ring can be obtained.

$$\frac{\hbar^2(\alpha^2 - n_\phi^2)}{r_3^2} + \frac{2m_o\hbar\alpha c}{r_3} + (m_o^2 - m_3^2)c^2 = 0 \qquad (7.21)$$

The solutions are

$$r_{3,\min} = \frac{\hbar n_\phi^2\left(1 - \frac{\alpha^2}{n_\phi^2}\right)}{m_o\alpha c\left(1 + \sqrt{1 - \left(\frac{n_\phi^2 - \alpha^2}{Y^2}\right)}\right)}$$

$$r_{3,\max} = \frac{\hbar n_\phi^2\left(1 - \frac{\alpha^2}{n_\phi^2}\right)}{m_o\alpha c\left(1 - \sqrt{1 - \left(\frac{n_\phi^2 - \alpha^2}{Y^2}\right)}\right)} \qquad (7.22)$$

where $Y = n_r + (n_\phi^2 - \alpha^2)^{1/2}$. Setting $n_r = 0$ in $Y$ for pure azimuthal states gives the circular ring radius

$$\bar{r}_3 = \frac{\hbar n_\phi^2\left(1 - \frac{\alpha^2}{n_\phi^2}\right)}{m_o\alpha c} \qquad (7.23)$$

This means that for a given $n_\phi$, with $n_r > 0$, we must have $r_{3,\min} < \bar{r}_3 < r_{3,\max}$. At this juncture, only the radial limits of the trajectory are determined. However, if we surmise an ellipse with semi-major and semi-minor axes $a$ and $b$, and eccentricity $\epsilon$, then $r_{3,\min} = a(1 - \epsilon)$ and $r_{3,\max} = a(1 + \epsilon)$. Since $b^2 = a^2(1 - \epsilon^2)$, the solutions arising from their average and product are

$$a = \frac{\hbar Y^2}{m_o\alpha c}$$

$$b = \frac{n_\phi\hbar Y\left(1 - \frac{\alpha^2}{n_\phi^2}\right)^{1/2}}{m_o\alpha c} \qquad (7.24)$$

The eccentricity is given by

$$\epsilon = \sqrt{1 - \frac{(n_\phi^2 - \alpha^2)}{Y^2}} \qquad (7.25)$$

The maximum and minimum radii of the elliptical OAM electron rings given by (7.22) are expressed in Table 7.2 as a fraction $\Gamma$ of the ground state radius $\bar{r}_{3,\mathrm{gs}}$, that is, of the circular $1S_{1/2}$ case in (7.23), in which $n_\phi = 1$, $n_r = 0$. As shown in Table 7.1, the Dirac theory distinguishes between degenerate levels by assigning a different spin value $\pm 1/2$ to each. However, in the MVR theory presented here, a spin quantum number will be superfluous. Two levels with identical $(n_\phi, n_r)$ are to share the same elliptic ring, and we shall see in Chapter 8 that what distinguishes them is their different fine-structure constant multipliers. In Table 7.2, $r_{3,\mathrm{min}} = \Gamma_{3,\mathrm{min}} \bar{r}_{\mathrm{gs}}$ and $r_{3,\mathrm{max}} = \Gamma_{3,\mathrm{max}} \bar{r}_{\mathrm{gs}}$, where $\bar{r}_{\mathrm{gs}}$ is the circular ground state radius of the electron.

Table 7.2   A selection of circular and elliptic fine-structure rings in the MVR theory of atomic hydrogen showing their minimum and maximum radii from (7.22) at *Kalpha* = 1.

| Level | $\Gamma_{3,\mathrm{min}}$ | $\Gamma_{3,\mathrm{max}}$ |
|---|---|---|
| $1S_{1/2}$ | 1 | 1 |
| $2S_{1/2}/2P_{1/2}$ | 0.535 897 281 2 | 7.464 315 734 2 |
| $2P_{3/2}$ | 4.000 159 762 6 | 4.000 159 762 6 |
| $3S_{1/2}/3P_{1/2}$ | 0.514 718 071 5 | 17.485 920 975 9 |
| $3P_{3/2}/3D_{3/2}$ | 2.291 885 866 0 | 15.708 912 946 7 |
| $3D_{5/2}$ | 9.000 426 033 5 | 9.000 426 033 5 |
| $4S_{1/2}/4P_{1/2}$ | 0.508 066 282 4 | 31.493 211 813 9 |
| $4P_{3/2}/4D_{3/2}$ | 2.143 678 052 2 | 29.857 813 064 8 |
| $5S_{1/2}/5P_{1/2}$ | 0.505 102 350 3 | 49.497 027 811 6 |
| $6S_{1/2}/6P_{1/2}$ | 0.503 521 143 | 71.499 674 10 |

*Note*: Each row represents one ring. Here, $r_{3,\mathrm{min}} = \Gamma_{3,\mathrm{min}} \bar{r}_{\mathrm{gs}}$ and $r_{3,\mathrm{max}} = \Gamma_{3,\mathrm{max}} \bar{r}_{\mathrm{gs}}$, where $\bar{r}_{\mathrm{gs}}$ is taken from (7.23) at $n_\phi = 1$.

### 7.2.7 Spin-2 and spin-3 radius

Using (6.9), with $p_{2f} = p_2$, $R_{2f} = r_2$, and (7.11), we have the invariant loaded spin-2 mass ring radius for all $(n_r, n_\phi)$ as follows:

$$r_2 = \frac{\hbar}{m_o c} \tag{7.26}$$

For the present, we shall assume the spin-2 profile to be circular, then in Chapter 9, we shall use an eccentricity $\epsilon < 1$ to obtain exact hyperfine energies. The generalized spin-2 angular momentum $J_2$ arises out of (7.11) and (7.26), while the generalized spin-3 angular momentum $J_3$ can be found using (7.12) and (7.23) for the circular $n_r = 0$ case,[8] so that

$$J_2 = p_2 r_2 = Z m_o c r_2$$
$$J_3 = \bar{p}_3 \bar{r}_3 = \frac{Z m_o \alpha c \bar{r}_3}{Y} \tag{7.27}$$

Here, the $Z$ is to be interpreted as the number of interlaced spin-2 loops in the SAM cluster. From (7.23) and (7.26), for the $n_r = 0$ case, we also find the relation of $\bar{r}_3$ to $r_2$ for a circular ring as follows:

$$\bar{r}_3 = \frac{n_\phi^2 \left(1 - \frac{\alpha^2}{n_\phi^2}\right)}{\alpha} r_2 \tag{7.28}$$

### 7.2.8 Spin-3 time period

For the loaded ring, let us consider only the case $n_r = 0$ so that $Y = (n_\phi^2 - \alpha^2)^{1/2}$. Using the spin-3 action derived from (7.27), (7.28) supported by (7.26), and the energy given by (7.14), the spin-3 time period for a field-free circular ring is

$$T_3 = \frac{2\pi J_3}{\varepsilon_3} = \frac{Z n_\phi^3 h \left(1 - \frac{\alpha^2}{n_\phi^2}\right)^{3/2}}{m_o \alpha^2 c^2} \tag{7.29}$$

---

[8]In Clarke (2017, 253–254), the proposal of a 'loaded' or 'radiation-affected' OAM ring is not addressed, but Principles 7.14 and 7.15 in that work give both $J_2$ and $J_3$ correctly as $\hbar$, for the $l = 1$ unloaded case.

## 7.3   The OAM Ring Field

### 7.3.1   *Field momentum map*

Let us summarize our scheme so far. We have already adopted the view that what we understand to be an observed localized photon is part of an array of parallel HSD (advancing spin-1 threads along guide tubes) in which one of the HSD excites a detector, and with smaller but non-zero chance, one or more others in the front might also produce an excitation. When spin-2 (optical OAM) is imposed on the HSD array, the spin-1 rays take a helical path round the optic axis accompanied by a field in the surrounding space, with a lower-bound radius of rotation $r_2$. The definition of mass is tied up with this radius $r_2$ or more precisely the perimeter; see (6.3). So, it can be said that SAM mass has already been created in the laboratory with the production of optical OAM. The optic axis is then bent into a closed circle to create an OAM mass ring at rest with spin-3 rotation. This produces two mutually orthogonal momentum components of the spin-1 ray: a spin-2 momentum $p_2$ perpendicular to the optic or ring tube axis as magnetic momentum, and a spin-3 momentum $p_3$ tangential to the OAM mass ring as electric momentum. The former is to be the direction of the magnetic field $\vec{B}$, while the electric field $\vec{E}$ is to be directed along the OAM ring axis; see Figure 7.1 for the proton ring.

As to how this spin-3 curvature is created, it is suggested here that as the electron OAM ring approaches the proton ring co-axially, the cumulative rotational effect of the increasing proton spin-3 field momentum in the electron ring displaces the oppositely rotating OAM of the SAM electron cluster into linear momentum. There comes a point when all the electron's OAM is converted into linear motion leaving no energy in the plane of the electron ring. Extending this analysis, if a linearly moving SAM cluster (optical OAM) is moving through a magnetic field (spin-3 field momentum) set perpendicular to it that increases in magnitude with distance traveled, the SAM cluster begins to adopt the rotation sense of the spin-3 magnetic field momentum. With sufficient field strength, such as might occur at atomic distances, an OAM mass ring might be created. This is explored in more detail in Chapter 8.

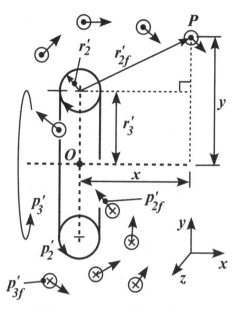

Figure 7.1   Momentum-field map at point $P(x, y, z)$ located on the electron OAM ring, for a cross-section of a proton OAM mass ring (primed variables) that lies in the $y - z$ plane. The ring has spin-2 radius $r_2'$ and spin-3 radius $r_3'$, while their respective momenta on the ring are $p_2'$ and $p_3'$. The spin-3 field momentum $p_{3f}'$ rises out of the page at the dotted circles, and runs into it at the crossed circles. The arrows on these circles are $p_{2f}'$ spin-2 field momenta, each with a length that represents the magnitude of the momentum which diminishes inversely as the spin-2 field radius $r_{2f}'$.

The spin-2 rest mass $m_{20}$, defined by (6.3), is equivalent to $m_0$ in the spin-3 motion given in (6.18). At higher radii than those on the OAM ring, there is also a rotation of HSD, and these constitute the magnetic and electric momentum fields, with components $p_{2f}'$ and $p_{3f}'$ for the proton (primed variables), respectively; see Figure 7.1. The mass scalar field is stretched out according to a scale factor that depends only on the radius $r_{2f}'$; see (6.9). The magnitude of the velocity vector field is invariant in that the resultant HSD speed is $c$ along all tubes, both in the ring and in the field.[9]

---

[9]In fact, apart from location, there need not be any distinction between the SAM ring and field which are both composed of rotating HSD spin-1 tubes. The ring exists at the minimum radius in the field.

In Maxwell's vortex scheme, there are contiguous vortices of matter rotating in the same sense which cause unwanted collisions from their surface counter rotations. To resolve this dilemma, he inserted idler particles between them, an invention that even he himself recognized as unrealistic (Maxwell 1861, Fig. 2, Plate VIII). In the present theory, at any location where two adjacent advancing screw threads are coincident, it is not mass particles that are in contact, but circularly polarized rotations that merely pass through each other.

The field radius $r'_{2f}$ always lies in a half-plane containing the OAM ring axis, that is, the $x$ axis; see Figure 7.1. As with optical OAM, we might imagine a void or singularity along this axis. At any point on this axis, if we take the components perpendicular to the $x$ axis of the two $p'_{2f}$ field momentum contributions, that rotate around diametrically opposed cross-sections of the tube, we find them to be in exact opposition. What remains are the components along the $x$ axis. So, it will be assumed here that inside the ring aperture, the fields from diametrically opposed tube cross-sections cannot affect each other. Anyway, at any moment, the SAM mass ring can only be in one place on the OAM ring. We now develop the equations that predict the map of the field momenta that surrounds the OAM spin-2 and spin-3 tube momenta.

There are two field maps as follows:

(I) The velocity field which everywhere has a spin-2 component given by (7.10) and a spin-3 component (7.9).

(II) The mass field map which, being dependent entirely on spin-2, analogous to (6.9) with (6.10), is derived from the division of $\hbar$ (spin-2 angular momentum) by the product of $v_2$ from (7.10) and $r_{2f}$.

Referring to Figure 6.9 (substituting a lower case radius for the loaded ring), let one extremity of the proton field radius $r'_{2f}$ fall on an electron ring at $A$, an OAM ring that is coaxial with the proton OAM ring. Then, if $r_2$ and $\bar{r}_3$ are the electron OAM mass ring's spin-2 and average spin-3 radii, respectively, and $\bar{r}'_3$ is the average spin-3 proton radius, if we take the angle that $r_2$ makes with the positive $z$ axis as $\gamma$, then since from (7.28) $r_2/\bar{r}_3 = \alpha$, we have

$$r'_{2f} = ((x + r_2 \cos\gamma)^2 + (\bar{r}_3 - \bar{r}'_3 + r_2 \sin\gamma)^2)^{\frac{1}{2}}$$

$$\sim (\bar{r}_3 - \bar{r}'_3)\left(1 + \frac{x^2}{(\bar{r}_3 - \bar{r}'_3)^2}\right)^{1/2} \tag{7.30}$$

Analogous to (6.9), and noting (7.27), the proton spin-2 field momentum passing through the electron's spin-3 circuit is then

$$p'_{2f} = \frac{Z'\hbar}{r'_{2f}} \tag{7.31}$$

Here, the primed variables refer to the proton, and $Z'$ is the number of interlaced units of spin-2 angular momentum. The spin-3 field momentum inhabiting the electron ring's spin-3 circuit in its ring plane arises from extending (7.13) to proton field momenta (primed variables) for circular states; thus,

$$p'_{3f} = \frac{Z'\hbar\alpha}{r'_{2f}Y'} \tag{7.32}$$

The assumption made here is that, similar to the electron, the proton has energy levels independently of an external field.[10]

### 7.3.2   *Coulomb's law: Proton field*

The calculation that takes place on the electron ring follows the scheme of Section 6.4.2. After replacing $R$ with $r$ throughout, we have the analogue of (6.25)

$$J'_{3f} = Z \int_{\gamma=0}^{\gamma=2\pi} \vec{p}'_{3f} \cdot d\vec{r}_2 \tag{7.33}$$

---

[10]For our hydrogen bound state, we place the proton in the ground state, where $n'_r = 0$, $n'_\phi = 1$, $Z' = 1$.

where the $Z$ indicates the number of interlaced spin-2 circuits. The spin-3 field momentum inhabiting the electron ring's spin-3 circuit arises from (7.13) in analogy with (6.9); thus,

$$\vec{p}_{3f}' = \frac{Z'\hbar\alpha}{r_{2f}'Y'} \begin{pmatrix} 0 \\ \cos\theta \\ -\sin\theta \end{pmatrix} \tag{7.34}$$

where $Z'$ is the number of units of angular momentum in the proton's spin-2 circuit, which is equivalent to the atomic number $Z'$ of the nucleus, and $Y' = n_\phi' + (n_r'^2 - \alpha^2)^{1/2}$ contains the proton quantum numbers.

Now, following the example of (6.24), we have

$$d\vec{r}_2 = r_2 d\gamma \begin{pmatrix} 0 \\ \cos\theta \\ -\sin\theta \end{pmatrix} \tag{7.35}$$

so, the proton-field action in the electron ring is

$$J_{3f}' = \frac{ZZ'\hbar\alpha r_2}{Y'} \int_0^{2\pi} \frac{d\gamma}{r_{2f}'} \tag{7.36}$$

Equation (7.10) applied to one circular spin-2 circuit ($n_r = 0$) of the electron SAM ring gives the spin-2 time period $T_2$ from

$$\frac{2\pi r_2}{T_2} = c\left(1 - \frac{\alpha^2}{n_\phi^2}\right)^{1/2} \tag{7.37}$$

So, dividing (7.36) by $T_2$, the electron spin-2 time period for circular states, we arrive at the proton-field energy inhabiting the electron ring, as follows:

$$\varepsilon_{3f}' = \frac{ZZ'\hbar\alpha c\left(1 - \frac{\alpha^2}{n_\phi^2}\right)^{1/2}}{2\pi Y'} \int_0^{2\pi} \frac{d\gamma}{r_{2f}'} \tag{7.38}$$

The time period $T_2$ is the time taken for a spin-1 ray (circularly polarized) to complete one spin-2 circuit (as optical OAM), yet in (7.33) there are $Z$ circuits. However, the $Z$ rays are to be interlaced around one circuit as suggested by Padgett and Allen (2000, 275, 278) and Padgett *et al.* (2004, Figure 1 caption),[11] then their transit time is only $T_2$.[12] Also, at any point in our derivation, we can substitute $\hbar a c = e^2/4\pi\epsilon_0$.

Now, from (7.26), we have

$$\hbar = m_o c r_2 \tag{7.39}$$

and from (7.28), we find the radial spin-2 to spin-3 ground state relation, $n_r = 0$, $n_\phi = 1$, as

$$\frac{r_2}{\bar{r}_{gs}} = \frac{\alpha}{1 - \alpha^2} \tag{7.40}$$

After using the relativistic mass $m$ from (7.6) with $n_r = 0$, for brevity we can write

$$\varepsilon'_{3f} = m\alpha^2 c^2 F'(x) \tag{7.41}$$

in which

$$F'(x) = \frac{ZZ'\bar{r}_{gs}\left(1 - \frac{\alpha^2}{n_\phi^2}\right)}{2\pi Y'(1 - \alpha^2)} \int_0^{2\pi} \frac{d\gamma}{r'_{2f}} \tag{7.42}$$

Here, the electron ground state radius $\bar{r}_{gs}$ is given by (7.23) at $n_\phi = 1$, $n_r = 0$. For a proton in ground state, $n'_\phi = 1$ and $n'_r = 1$ in $Y'$, and we have

$$F'(x) = \frac{ZZ'\bar{r}_{gs}\left(1 - \frac{\alpha^2}{n_\phi^2}\right)}{2\pi\left(1 - \alpha^2\right)^{3/2}} \int_0^{2\pi} \frac{d\gamma}{r'_{2f}} \tag{7.43}$$

---

[11]These authors refer to "$l$ [$=Z$] intertwined fronts" and "the number of $l$ [$=Z$] intertwined helical phase fronts", respectively.

[12]In fact, if the circuits were not interlaced but were longitudinally iterated instead, then we would need a time $ZT_2$ and the Coulomb law would not result.

Here, the prime on the $F'(x)$ refers to the 'proton'. We should emphasize here that (7.43) only applies to circular states of the electron.

### 7.3.3   Coulomb's law: Electron field

The action calculation for the effect of the electron field on the proton ring is more or less symmetrical, so that similar to (7.33), we have

$$J_{3f} = Z' \int_{\gamma=0}^{\gamma=2\pi} \vec{p}_{3f} \cdot d\vec{r}_2' \tag{7.44}$$

Following (7.34), the electron field momentum inhabiting the proton ring is

$$\vec{p}_{3f} = \frac{Z\hbar\alpha}{r_{2f}Y} \begin{pmatrix} 0 \\ -\cos\theta \\ \sin\theta \end{pmatrix} \tag{7.45}$$

where $Z$ is the number of interlaced electron spin-2 circuits. The analogue of (7.35) is

$$d\vec{r}_2' = r_2' d\gamma \begin{pmatrix} 0 \\ \cos\theta \\ -\sin\theta \end{pmatrix} \tag{7.46}$$

so that

$$J_{3f} = \frac{ZZ'\hbar\alpha r_2'}{Y} \int_0^{2\pi} \frac{d\gamma}{r_{2f}} \tag{7.47}$$

Here, $r_{2f} = r_{2f}'$ as given in (7.30). Equation (7.10) taken on the circular proton ring gives

$$\frac{2\pi r_2'}{T_2'} = c \left( 1 - \frac{\alpha^2}{n_\phi'^2} \right)^{1/2} \tag{7.48}$$

to give the proton spin-2 time period $T_2'$ for one circuit. So, dividing (7.47) by $T_2'$, we arrive at the electron-field energy permeating the proton ring, as follows:

$$\varepsilon_{3f} = \frac{ZZ'\hbar\alpha c \left(1 - \frac{\alpha^2}{n_\phi'^2}\right)^{1/2}}{2\pi Y} \int_0^{2\pi} \frac{d\gamma}{r_{2f}} \qquad (7.49)$$

Using (7.39) and (7.40), we have

$$\varepsilon_{3f} = m\alpha^2 c^2 F(x) \qquad (7.50)$$

where

$$F(x) = \frac{ZZ'\bar{r}_{gs} \left(1 - \frac{\alpha^2}{n_\phi'^2}\right)^{1/2}}{2\pi n_\phi \left(1 - \alpha^2\right)} \int_0^{2\pi} \frac{d\gamma}{r_{2f}} \qquad (7.51)$$

having used $Y = n_\phi(1 - \alpha^2/n_\phi^2)^{1/2}$ for circular states. With the proton in its ground state, we have $n_\phi' = 1$ and $n_r' = 0$, so this then becomes

$$F(x) = \frac{ZZ'\bar{r}_{gs}}{2\pi n_\phi(1 - \alpha^2)^{1/2}} \int_0^{2\pi} \frac{d\gamma}{r_{2f}} \qquad (7.52)$$

We note that (7.43) is the same as (7.52) when $n_\phi = 1$. In other words, the field effect of the proton on the electron and vice versa is the same when the electron is in ground state.

It Chapter 8, we shall set up the proton–electron bound state as a hydrogen atom. The idea will be developed that when the spin-3 proton field energy, $\varepsilon_1$ say, occupying the electron ring displaces an equal amount of spin-3 electron ring energy $\varepsilon_1$ into linear motion along the $x$-axis, an amount $\varepsilon_2$ remains in the electron ring. When $\varepsilon_1 = \varepsilon_2$, which corresponds to equal field and electron spin-3 momenta in the electron ring plane, the energy of electron motion along the $x$ axis is radiated away and the electron OAM ring stops. This radiation occurs when the electron and proton are located at the entrance to a *bound state*, and thereafter the electron ring oscillates coaxially over the smaller proton ring.

### 7.3.4  *Integral computation*

Here, we obtain the integral in (7.43) and (7.52) to fourth order in $\alpha$, noting that $r'_{2f} = r_{2f}$. We can prepare the ground by rewriting (7.30) as

$$
\frac{1}{r'_{2f}} = \frac{1}{\bar{r}_3(1 - R_1)\left(1 + \frac{X^2 + R_2^2}{(1 - R_1)^2}\right)^{1/2}}
$$

$$
\times \left(1 + \frac{2R_2}{(1 - R_1)^2} \frac{(X \cos\gamma + (1 - R_1)\sin\gamma)}{\left(1 + \frac{X^2 + R_2^2}{(1 - R_1)^2}\right)}\right)^{-1/2} \tag{7.53}
$$

We should note that the $R_1$ and $R_2$ are no longer unloaded-ring radii and have been enlisted here for a new role. Also, $\bar{r}_3$ is for a general circular state characterized by $n_\phi$ and using (7.28), we have

$$
X = \frac{x}{\bar{r}_3} = \frac{x}{\Gamma \bar{r}_{gs}}
$$

$$
R_1 = \frac{\bar{r}'_3}{\bar{r}_3} = \frac{\bar{r}'_3}{\Gamma \bar{r}_{gs}} = \frac{m_o}{\Gamma m'_o} \tag{7.54}
$$

$$
R_2 = \frac{r_2}{\bar{r}_3} = \frac{r_2}{\Gamma \bar{r}_{gs}} = \frac{\alpha}{\Gamma(1 - \alpha^2)^{1/2}}
$$

Here, we note that $\bar{r}'_3/\bar{r}_{gs} = m_o/m'_o$ from (A.15), and $r_2/\bar{r}_{gs} = \alpha/(1 - \alpha^2)^{1/2}$ from (A.3). It should be remarked that we can cancel out the spin-3 ground state radius $\bar{r}_{gs}$ in (7.42) and (7.52) having made the substitution $\bar{r}_3 = \Gamma \bar{r}_{gs}$ when using (7.54) to produce (7.55), given that $\Gamma$ is the fraction of the spin-3 ground-state radius taken from Table 7.2. Here, $\Gamma$ can represent a circular radius or a perigee or an apogee from an elliptic ring, and $X$ becomes the axial proton–electron separation distance, both expressed as a fraction of the average electron spin-3 ground state radius $\bar{r}_{gs}$.

The first task is to execute the integral over $\gamma$ to fourth order in the binomial series. Thus,

$$f(x) = \int_0^{2\pi} \frac{1}{r'_{2f}} d\gamma = \frac{2\pi}{\bar{r}_3(1 - R_1)\left(1 + \frac{X^2 + R_2^2}{(1-R_1)^2}\right)^{1/2}}$$

$$\times \left(1 + \frac{3R_2^2(X^2 + (1 - R_1)^2)}{4(1 - R_1)^4\left(1 + \frac{X^2 + R_2^2}{(1-R_1)^2}\right)^2}\right.$$

$$+ \left.\frac{105R_2^4(X^2 + (1 - R_1)^2)^2}{64(1 - R_1)^8\left(1 + \frac{X^2 + R_2^2}{(1-R_1)^2}\right)^4} + \cdots\right) \tag{7.55}$$

For an emission to occur from the electron ring, we demand that the proton field energy in the electron ring energy is half the total electron ring energy. This equality is given by the equivalence of (7.17) and (7.41)–(7.43), so that

$$F' = \frac{1}{2(Y^2 + \alpha^2)} \tag{7.56}$$

and so we need to conduct a search for the $X = x/\bar{r}_{gs}$ that satisfies (7.56). This is the separation of the proton and electron ring planes taken along their common axis, and expressed as a fraction of the average electron spin-3 ground state radius $\bar{r}_{gs}$. This is carried out in the [*coordinates*] subroutine (lines 600–642, specifically line 627) in the computer program in Appendix C.

## References

Clarke, B. R. *The Quantum Puzzle: Critique of Quantum Theory and Electrodynamics*. Singapore: World Scientific Publishing, 2017.

Horbatsch, M., and E. A. Hessels. 'Tabulation of the bound-state energies of atomic hydrogen'. *Physical Review A*, 93 (2016): 022513.

Kramida, A. E. 'A critical comparison of experimental data on spectral lines and energy levels of hydrogen, deuterium, and tritium'. *Atomic Data and Nuclear Tables*, 96 (2010): 586–644.

Maxwell, J. Clerk. 'On physical lines of force'. *Philosophical Magazine*, Part 1, xxi (1861): 161–175.

Padgett, M., and L. Allen. 'Light with a twist in its tail'. *Contemporary Physics*, 41 (2000): 275–285.

Padgett, M., J. Courtial, and L. Allen. 'Light's orbital angular momentum'. *Physics Today*, May (2004): 35–40.

Sommerfeld, A. *Atomic Structure and Spectral Lines*. Translated from the third German edition by Henry L. Brose. London: Methuen, 1923.

White, H. E. *Introduction to Atomic Spectra*. New York: McGraw-Hill, 1934.

Chapter 8

# Hydrogen Atom Fine Structure

*The hydrogen bound state is set up and the ionization energy is calculated. It is postulated that all emissions and absorptions occur at the oscillation boundary of the $1S_{1/2}$ circular ring, and all excited rings collapse to the $1S_{1/2}$ ring for transitions. At the boundary, the proton field energy and the remaining electron energy in the electron ring are equal. All transitions are subject to a Doppler shift in which the proton speed of approach is taken into account. A detailed analysis of the computer program used for calculating transition frequencies and emission coordinates is given. The electron and proton oscillations are $\pi/2$ out of phase and the mechanism is examined in detail. Hints for electron configuration are given based on the Sommerfeld quantum numbers and the two spin-2 rotation senses.*

## 8.1  Overview

### 8.1.1  *History*

The English mathematician John Nicholson was the first to suggest the quantization of angular momentum in atomic systems. His model assumed $n$ equidistant electrons placed around a single circular orbit with a central nucleus, and required that "the angular momentum of an atom can only rise or fall by discrete amounts involving Planck's constant $h$ when electrons leave or return" (Nicholson 1912, 679). Niels Bohr (1913a) extended Nicholson's single-orbit atom to a system with several stable non-radiating orbits, contradicting classical electromagnetic theory which predicted that the electron should accelerate into the nucleus, continuously radiating away its energy. He found that the energy differences between the levels could account reasonably well for the Balmer and Paschen series for hydrogen. By following an analogy with Keplerian planetary motion, Bohr (1913b) conceived of the atomic nucleus and the electron revolving about their center of mass. The resulting reduced mass correction factor not only improved the accuracy of his earlier calculations but also allowed him to extend his predictions to a single-electron helium atom.

Bohr's system depended on only a single quantum number $n$. However, as early as 1887, Albert Michelson had observed that "the red hydrogen line $[H_\alpha]$ must be a double line" (Michelson 1887, 466). This raised the possibility that an extra quantum number would be needed. Two years before Bohr's paper, at the 1911 Solvay Conference, Henri Poincaré had enquired about generalizing a quantized system to several degrees of freedom (Mehra and Rechenberg 2001, 208). The task was taken up by William Wilson who subsequently gave an action integral for each of three quantum numbers (Wilson 1915, 796), and later gave the quantum conditions for elliptic orbits of an electron in polar coordinates, with its eccentricity given as a function of the azimuthal and radial quantum numbers $n_\phi$ and $n_r$ (Wilson 1916).[1] In the same year, publishing later than Wilson, Arnold Sommerfeld also gave the eccentricity function, but extended the investigation by producing the relativistic energy of an electron in a hydrogen atom (Sommerfeld 1916; 1923, 472–473); see Equation (B.4).[2] On the basis of Heisenberg's quantum mechanics, after significant contributions by Pauli (1926) and Thomas (1926), Heisenberg and Jordan (1926) obtained the same fine-structure formula using Kronig's spin quantum number.[3]

---

[1] The idea of fine-structure doublets being due to the relativistic mass variation of the electron in an elliptic orbit had been suggested by Bohr (1915). However, he abandoned the idea due his perceived inaccuracy of the measurements.

[2] Mehra and Rechenberg (2001, 215) report that Sommerfeld had arrived at Wilson's results "independently and earlier".

[3] For a discussion of the introduction of the spin quantum number, see (Kragh 1985, 114). Following Kronig's suggestion, Goudsmit and Uhlenbeck (1925) relabeled the energy levels $(n, k, j)$, where $k = l + 1$ and $j = l \pm 1/2$. They introduced the spinning electron a year later (Uhlenbeck and Goudsmit 1926).

### 8.1.2   *The new MVR theory*

There are several problems with the present state of the theory of the hydrogen atom structure. First, both the Sommerfeld and quantum mechanics derivations of the fine-structure formula depend on an external potential to obtain the energy levels. Unfortunately, it is only an *ad hoc* term, because over 230 years after its introduction by Laplace, its internal structure is still in need of elucidation[4]:

> We have theories relating to these [E-M] fields, but we have no idea whatever of what they are intrinsically, nor even the slightest idea of the path to follow in order to discover their true nature. (Bjerknes 1906, 1)

This is not the only problem with electric potential. As stated in Chapter 6, recent evidence suggests that an electron vortex can adopt energy levels in the absence of an external potential:

> Electrons can be prepared in quantized orbital states with large OAM, in free space devoid of any central potential, electromagnetic field, or medium that confines the orbits.[5] (McMorran *et al.* 2011, 194)

The last visualizable theory of the hydrogen atom was Sommerfeld's elliptic orbits, which is over 100 years old, and since that time, no pictorial theory that accounts for the hyperfine splitting has been presented. In fact, Dirac (2000, 10) even suggested abandoning such attempts:

> the main object of science is not the provision of pictures, but it is the formulation of laws governing phenomena and the application of these laws to the discovery of new phenomena [. . .] In the case of atomic phenomena no picture can be expected to exist in the usual sense of the word 'picture' by means of which is meant a model functioning essentially on classical lines.

---

[4]The work of James MacCullagh (1880) from December 1839 on a rotationally elastic ether should not be overlooked here, although he had no visualizable model.
[5]See also Molina–Terriza *et al.* (2001).

Finally, a theory is still needed that reduces the concepts of mass and charge to a unified scheme, one based on a clear geometrical model and its mode of operation. With these objections in view, an MVR theory of the hydrogen atom will now be presented. The fundamental element out of which the whole edifice is constructed is that which has been developed in earlier chapters: the circularly polarized ray or helical space dislocation (HSD) or spin-1 rotation.[6] In the atomic model presented here, it will be set in motion along the surface of a torus. Its passage around the toroidal cross-section will constitute an SAM mass ring (magnetic momentum), and its motion around the axis of the toroid will give the OAM mass ring (electric momentum).

Consider an electron OAM ring and a proton OAM ring, where the latter is simply a scaled down replica of the former.[7] Let the electron ring approach the proton ring coaxially from an arbitrarily large distance $x \to \infty$; see Figure 8.1. The proton electric momentum field (crossed and dotted circles in Figure 7.1) intruding on the electron OAM ring increases as the separation of the rings decreases. Consequently, the field energy permeating the electron ring results in the displacement of an equal amount of electron energy from its ring into linear energy along the $x$ axis, conserving the total electron energy in the ring. The SAM trajectory is no longer confined to the electron ring but is deflected at an angle $\phi$ to it. As the proton–electron separation distance decreases, the SAM cluster displacement increases, the electron ring linear speed increases, and $\phi$ increases until the entrance to the proton–electron bound state $B$ is reached; see Figure 8.1. At this point, we find that the intruding field energy in the electron ring and the remaining electron spin-3 energy are equal at $m\alpha^2 c^2/2$ each. The electron OAM ring now radiates away its linear energy of motion as ionization energy, as the ring comes to rest at the bound-state entrance, leaving an amount of energy

---

[6]We recall the distinction between them: the HSD is a traveling non-rotating screw thread that creates a circular polarization or spin–1 rotation as it passes through a stationary plane set perpendicular to the thread axis.

[7]The length scale factor is $m/m'$ , where $m$ and $m'$ are the electron and proton masses, respectively.

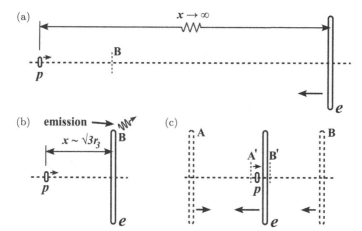

Figure 8.1   Side view of the construction of the $1S_{1/2}$ bound state of a hydrogen atom. (a) Co-axial arrangement of a proton OAM ring p (left) and an electron OAM ring e (right) which are initially set a distance $x \to \infty$ apart. The electron and proton approach the electron's entrance to the bound state at $B$ at which point the electron emits. (b) At B, the field energy is $m\alpha^2 c^2/2$, and we find that $x \sim \sqrt{3}\bar{r}_3$. As shown, the electron emits energy away from p in a Doppler blue shift (toward for a red shift) and the electron is brought to rest at B. The proton does not emit and its motion continues toward $B$ unaffected. (c) The electron ring slowly accelerates left away from B toward the proton and oscillates between $A$ and $B$, while the proton ring oscillates between $A'$ and $B'$ (not to scale), $\pi/2$ out of phase (see Section 8.4).

circulating around the ring equal to that radiated given by (7.17). It is also equal to the displacing energy of the proton field in the electron ring plane.

As far as the fine structure is concerned, two further considerations must be introduced to (7.17): the reduced mass of the electron, which is treated in Section 6.5.2; and the innovation of a Doppler shift (with respect to the proton rest frame) of the radiation emitted at the bound-state entrance. Both are transformations to the proton rest frame. For the Doppler shift, the angle of emission of the radiation between a point on the electron ground state circular ring, a point at the proton ring origin, and a point at the electron ring origin must be considered; see Figure 8.3. The velocity component of the advancing proton with respect to the emission direction can then be calculated. Now, in order to emit (and absorb), it is posited that all levels collapse to the circular ground state OAM ring at the bound-state entrance. So, to calculate the above-mentioned angle of emission, the spin-3 radius of the $1S_{1/2}$ electron ring is calculated (*gamma* in the computer program) together with the proton-electron ring separation distance (*Xstore*). A red shift occurs when the emission exits the bound state through the receding proton ring, and a blue shift when it runs away from an advancing ring.

Certain fine-structure levels share the same elliptical ring: $2S_{1/2}$ and $2P_{1/2}$, $3S_{1/2}$ and $3P_{1/2}$, $3P_{3/2}$ and $3D_{3/2}$; see Table 7.2. In MVR theory, the degeneracy of the fine structure is removed, not by making use of a spin of $\pm 1/2$, which will be redundant here, but by positing two different values of the fine-structure constant multiplier *Kalpha*. The value of *Kalpha* is obtained through a computer search under the condition that, for the level in view, the fine-structure energy is brought into alignment with the mid-point of the two experimentally determined hyperfine frequencies (MHz). In Chapter 9, the spin-2 circuit of the electron will be given an eccentricity different from zero so that the two spin-2 rotation senses can produce a raising or lowering of this hyperfine mid-point to coincide with the appropriate experimental hyperfine value. This is carried out using a computer search for the eccentricity.

This is a general outline of how the present MVR hydrogen atom structure is to function, a model that will now be set out in more detail.

## 8.2   Setting up the Hydrogen Bound State

### 8.2.1   *The ionization energy*

We first set up a bound state for the proton and electron. A field-free state may be constructed for any loaded ring, with quantum numbers $(n_\phi, n_r)$, with the field-free energies given by (7.14), slightly modified here as (8.1)[8]

$$\mathcal{E} = m \frac{\alpha^2 c^2}{Y^2 \left(1 + \frac{\alpha^2}{Y^2}\right)} = m\alpha^2 c^2 \tag{8.1}$$

---

[8]The electron rest mass $m_o$ is replaced by $m'_o$ when we consider the proton case.

Here, the relativistic electron mass $m$ is suggested by (7.11) and $Y = n_r + (n_\phi^2 - \alpha^2)^{1/2}$.[9]

Let us permit the axes of the proton and electron to be aligned, with their distance apart set arbitrarily large; see Figure 8.1(a).[10] This condition of alignment, where the proton axis penetrates the electron OAM ring aperture, is necessary for the electron to be affected by the proton's electric field.[11] Without it, there can be no electrical interaction, only a spin-2 magnetic one. The electron ring is now advanced an arbitrarily small distance (to the left in Figure 8.1) toward the proton ring. This reduction in the separation distance results in two changes: an increase (from zero if $x \to \infty$) in spin-3 proton field momentum inhabiting the electron ring, and an increase (from zero if $x \to \infty$) in the spin-3 electron momentum field permeating the proton ring.[12] In consequence, if the rings exhibit opposing spin-3 rotation senses, the field energy intruding in each ring redistributes the equivalent ring energy into linear energy of the ring motion along the common axis of the interacting rings. Consequently, they accelerate toward each other in an electrical attraction.[13]

As pointed out in Section 6.4.2.1, the redistribution of the host ring energy due to its motion does not diminish the field it produces. For the attraction case, the sum of the remaining energy and the external field energy in the host ring plane is constant. This is the basis for the host ring's invariant field effect.[14]

If we confine our attention to the ground states of the proton and electron, then for the electron $1S_{1/2}$ state, we have $n_\phi = 1$, $n_r = 0$. According to (6.47), we can replace $m_o$ in (8.1) with the

---

[9]With a slight change of notation, use is made of the Sommerfeld quantum numbers: radial $n_r$ and azimuthal $n_\phi$.

[10]Their common axes need not even be linear. So long as the axis of the proton penetrates the electron ring aperture, there is a Coulomb effect.

[11]This was first pointed out in Clarke (2017, 313). The electron ring axis must also penetrate the proton ring aperture for a reciprocal effect.

[12]This is the electron spin-3 field momentum $p_{3f}$ in Figure 6.5(b).

[13]Same-sense rotations produce repulsion.

[14]This is analogous to $P.E. + K.E. = $ constant. It is not entirely kinetic energy in the electron ring because the ring energy includes its self-potential.

reduced mass $\mu \sim m_o/(1 + m_o/m'_o)$.[15] This transforms the electron mass from the center of mass frame to the proton rest frame. With these restrictions, (8.1) now becomes

$$\mathcal{E} = \mu \frac{\alpha^2 c^2}{(1 - \alpha^2)^{1/2}} \tag{8.2}$$

### 8.2.2　*The proton field energy*

We now consider the proton field energy in the electron OAM ring. For ease of illustration, we refer to (7.43), the field from the proton in ground state, and make a number of approximations. Referring to Figure 8.1, the electron increasingly takes on proton field energy as it approaches the proton and, in order to maintain the total energy in the electron ring (proton field energy plus remaining electron ring energy), the proton field displaces an equivalent electron ring energy into linear momentum of the OAM ring toward the proton along their common axis. For the electron in ground state, from (7.41), (7.43), (7.54), and (7.55), with $Z = Z' = 1$, we find this field energy to be

$$\varepsilon'_{3f} \sim \mu \frac{\alpha^2 c^2}{(1 - \alpha^2)^{1/2} \left(1 + \frac{x^2}{\bar{r}_{gs}^2}\right)^{1/2}} \tag{8.3}$$

Since we are now considering electron mass and radiation emission with respect to the proton rest frame, to arrive at (8.3), we use a reduced mass $\mu$. In (7.43) we have also taken $n_\phi = 1$, while in (7.54), we have $\Gamma = 1$ for the electron ground state radius.

### 8.2.3　*Energy equivalence at boundary*

We now return to the electron OAM ring energy and demand that the spin-3 momentum transformation given by (7.16) applies to (8.2), so that $\alpha \to \alpha/\sqrt{2}$; see Figure 8.2.

---

[15]In what follows, the electron rest mass in the denominator of the reduced mass in (6.47) will be replaced by its relativistic equivalent (7.17) at the proton–electron oscillation boundary.

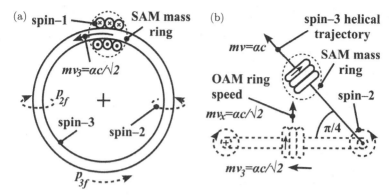

Figure 8.2   Electron OAM mass ring in ground state at the entrance to a proton–electron bound state. (a) The SAM mass ring cluster moves into the page at speed $\alpha c/\sqrt{2}$, and moves around the OAM ring at spin-3 speed $v_3 = \alpha c/\sqrt{2}$. (b) Plan view showing the spin-3 velocity component $v_3 = \alpha c/\sqrt{2}$ (horizontal arrow) in the ring plane as the electron SAM mass ring moves perpendicular to the ring plane (vertical arrow) at the OAM ring linear speed $v_x = \alpha c/\sqrt{2}$. This causes the cluster to advance at the invariant speed $v = \alpha c$ on a helical trajectory at a rake angle of $\pi/4$ relative to the electron ring plane.

Equation (8.2) for the resident electron ring energy is therefore modified to

$$\mathcal{E}_{\text{res}} = \frac{\mu}{\left(1 - \frac{\alpha^2}{2}\right)^{1/2}} \frac{\alpha^2 c^2}{2} \tag{8.4}$$

This is also equivalent to the linear energy of the electron OAM ring at its point of emission at $B$ in Figure 8.1. At the oscillation boundary, the electron ring is brought to rest as a consequence of radiating away this energy of motion.[16] It is the ionization energy that is to be supplied to release the electron OAM ring intact from the bound state.[17] Traditional theory informs us that the process of ionization involves the electron (here, the SAM cluster) being raised

---

[16]The proton OAM ring, since it is still inhabited by an unchanged electron field energy originating from the electron ring plane, continues its motion toward the electron ring. It does not emit since its velocity is far below that needed for a $\pi/4$ rake angle in its SAM trajectory.

[17]We also need to take account of the Doppler blue shift and the hyperfine adjustment.

to greater and greater ring radii until it leaves the field of influence of the proton. This is not the mechanism proposed here, although the ionization energy obtained would be the same. In MVR theory, the electron always stays in an OAM ring and we simply reverse the process of emission in the ground state, that formed the bond with the proton, by resupplying the radiated energy. This restores the energy of motion of the OAM ring and allows it to escape to $x \to \infty$. In other words, (7.17) reverts to (7.14).

In Figure 8.2, the rake angle of the helical trajectory is $\pi/4$ due to the linear ring momentum being $mv_x = m\alpha c/\sqrt{2}$ and the spin-3 helical trajectory momentum being $mv = m\alpha c$. Now, to transform from momentum to energy for spin-3 on a stationary OAM mass ring, we multiply by $\alpha c$. However, the relativistic effect due to motion means that this becomes $\alpha c/\sqrt{2}$, so we can see that the electron energy remaining in the ring plane is the difference between that on its helical trajectory and its linear energy, that is, $m\alpha^2 c^2 - m\alpha^2 c^2/2 = m\alpha^2 c^2/2$.

Since, at the oscillation boundary of the hydrogen atom, the proton field energy in the electron ring is the same as the electron energy remaining in its ring, we ask what value of $x$ accrues if we set (8.3) equal to (8.4). This amounts to solving the approximation

$$\left(1 + \frac{x^2}{\bar{r}_{gs}^2}\right)^{1/2} \sim 2 \tag{8.5}$$

where $\bar{r}_{gs}$ is the average spin-3 ground state radius and we find

$$x \sim \sqrt{3}\bar{r}_{gs} \tag{8.6}$$

We must remind ourselves that the proton field energy (8.3) is only an approximation to that obtained from (7.55) and (7.56). Nevertheless, we can state that the coordinates of the point of emission on the electron circular ring with respect to the proton ring center are approximately $(x, y) = (\sqrt{3}, 1)$, given as multipliers of $\bar{r}_{gs}$.

### 8.2.4 *Doppler shift adjustment*

The notion of a reduced mass transforms the electron mass from the center of the mass frame to the proton rest frame. However, the emitted radiation must also be referred to the proton frame as a Doppler shift. With the proton in motion, it is as if its field, the solid ground upon which the electron moves, is moving beneath the electron ring, and all calculations must be related to a proton field treated as if it is at rest (assuming the emission is not perpendicular to the proton motion). The $x$ coordinate of the emission is calculated from the accurate condition (7.56) that the proton field energy in the electron OAM ring equals the energy remaining in the electron ring. The $y$ coordinate is taken from (7.22) for an elliptical ring or (7.23) for a circular ring. We now construct a Doppler term to modify the fine-structure energy calculation (7.17) for $1S_{1/2}$ as the emission travels away from the bound system. In Figure 8.3, when the emission is away from the proton, the proton has a component of velocity toward the emission point and there must be a blue shift. However, the emission might instead pass through the proton so that on emergence the proton is receding from it to produce a red shift.

Let the location of the emission $(x, y)$ with respect to the proton spin-2 axis, which is where its field originated from, determine the

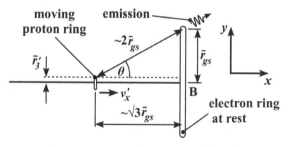

Figure 8.3   Side elevation of the ground state electron emission at the entrance B to the bound state. The electron ring enters from the right and comes to rest at B in the center of mass frame with an emission, but the non-emitting proton ring still moves at speed $v'_x$ toward it. The blue-shifted emission (shown) runs parallel to the spin-2 field radius of the proton, away from and at an angle $\theta$ to the advancing proton. Alternatively, a red-shifted emission passes in the opposite direction and emerges on the other side of a receding proton.

angle $\theta$ (see Figure 8.3) as follows:

$$\cos \theta = \frac{|x|}{(x^2 + (y - \bar{r}'_{\mathrm{gs}})^2)^{1/2}} \tag{8.7}$$

where, from the loaded-ring version of (A.15), the proton spin-3 radius $\bar{r}'_{\mathrm{gs}} = m_o \bar{r}_{\mathrm{gs}}/m'_o = M\bar{r}_{\mathrm{gs}}$.[18] Also, $(x/\bar{r}_{\mathrm{gs}}, y/\bar{r}_{\mathrm{gs}})$ are coordinates with respect to the proton OAM ring center, given as multipliers of the electron ring's spin-3 ground state radius $\bar{r}_{\mathrm{gs}}$. In Figure 8.3, these coordinates are approximately $(\sqrt{3}, 1)$.[19] The proton speed $v'_x$ is given by a conservation of linear momentum in the center of mass frame, see (6.40), so that

$$\frac{v'_x}{c} = \frac{m}{m'}\frac{v_x}{c} = \frac{\alpha M}{\sqrt{2}} \tag{8.8}$$

where $M = m_o/m'_o$, so after use of (8.7) and (8.8), the fine-structure ground state term (8.4) becomes Doppler-adjusted to

$$\mathcal{E}_1 = \frac{m}{\left(1 + \frac{m}{m'_o}\right)} \frac{\alpha^2 c^2}{2\left(1 - \frac{\alpha^2}{2}\right)^{1/2}} \left(1 + \frac{v'_x}{c}\cos\theta\right)$$

$$\sim \frac{m}{\left(1 + \frac{m}{m'_o}\right)} \frac{\alpha^2 c^2}{2\left(1 - \frac{\alpha^2}{2}\right)^{1/2}} \left(1 \pm \frac{\alpha M}{\sqrt{2}}\frac{\sqrt{3}}{(3 + (1 - M)^2)^{1/2}}\right) \tag{8.9}$$

Here, the $+$ sign in the final term produces a blue shift and the $-$ sign a red shift. We note that, in the reduced mass, the electron rest mass $m_o$ has been replaced by the relativistic electron mass $m = m_o/(1 - \alpha^2/2)^{1/2}$ for greater accuracy.

The computer program used to investigate the hydrogen atom is given in Appendix C. To obtain the ionization frequency (MHz), that is, the $1S_{1/2}$ emission, we run the program and input '$1S_{1/2}$' when prompted.[20] The results are shown in Table 8.1.

---

[18] Here, $\bar{r}_{\mathrm{gs}} = r_3$ is the loaded ring analogue of the unloaded $R_3$, and $m_o = m_{2o}$.
[19] These coordinates are in need of refinement as condition (7.56) has only been approximated here for brevity.
[20] For the other three input values, '$+$', '135', and 'blue' will suffice, omitting the quotes.

Table 8.1 Results for the fine-structure $1S_{1/2}$ ionization frequency emission (MHz) for hydrogen. Comparison of Sommerfeld, MVR, Doppler-adjusted MVR, and the experimental frequency (Horbatsch and Hessels 2016, Table III) using the CODATA (Mohr *et al.* 2017) value of the fine-structure constant (at *Kalpha* = 1). The other three computer program inputs are '+', '135', and 'blue'.

| Key | Definition | Value |
|-----|------------|-------|
| (a) | Expt hyperfine average | 3 288 087 212.2131 |
| (b) | Sommerfeld f-s | 3 288 095 006.0264 |
| (c) | MVR f-s (pre-Doppler, *Kalpha* = 1) | 3 288 094 981.9247 |
| (d) | MVR *f-s* (pre-Doppler, adjusted *Kalpha*) | 3 288 079 208.3160 |
| (e) | MVR *f-s* (post-Doppler, adjusted *Kalpha*) | 3 288 087 212.2131 |
| (f) | Doppler factor (circle) | 1.000 002 434 216 6310 |
| (g) | *Kalpha*($\alpha_1$) | 0.999 997 601 435 212 6 |
| (h) | $x/\bar{r}_{gs}$ | 1.732 412 679 2 |
| (i) | $y/\bar{r}_{gs}$ | 1.0 |

In Table 8.1, we find the following for the ground state computer program output (see Appendix D.2):

(a) The average of the two experimentally measured hyperfine frequencies (MHz) of $1S_{1/2}$ (Horbatsch and Hessels 2016, Table III) — computer output value (15).

(b) The Sommerfeld fine-structure frequency using the accepted fine-structure constant (*Kalpha* = 1) — output value (1).

(c) The MVR fine-structure frequency without Doppler shift using the accepted fine-structure constant (*Kalpha* = 1) — output value (2).

(d) The MVR fine-structure frequency without Doppler shift using the adjusted fine-structure constant — output value (12).

(e) The MVR fine-structure frequency with Doppler shift using the adjusted fine-structure constant chosen to align with (a) — output values (14) and (20).

(f) The Doppler blue-shift value that multiplies (d) based on the angle of emission relative to the advancing proton velocity — output value (10).

    (g) The decimal-place searched fine-structure constant multiplier *Kalpha*, determined by the condition that (e) aligns with (a) — output value (24).

(h) & (i) The proton-electron separation $x/\bar{r}_{gs} = Xstore$ and the electron ring radius $y/\bar{r}_{gs} = gamma$ based on the adjusted fine-structure constant, expressed as multipliers of the ground state radius $\bar{r}_{gs}$ — output values (6) and (7).

The fine-structure constant $\alpha$ is automatically adjusted by varying *Kalpha* in $\alpha_1 = Kalpha \times \alpha$.

When there is a blue shift, the set of *Kalpha* $< 1$, but with a red shift *Kalpha* $> 1$. The electron spin-2 eccentricities are essentially the same.

## 8.3    The Fine-Structure Calculation

### 8.3.1    *Preliminary*

Section 8.2 has given a sketch of how the calculation operates for the hydrogen fine-structure ionization energy, by considering the Doppler-shifted radiation frequency from the $1S_{1/2}$ state. We shall now analyze the computer program in Appendix C in an attempt to remove any doubt as to how the scheme operates. Here, we shall address the four inputs, the automatic input checks, the given data, the Sommerfeld calculation, the new MVR fine-structure calculation (7.17) with reduced mass, the calculation of the emission coordinates on the OAM ring, and the construction of the Doppler blue-shift multiplier that modifies the fine-structure frequency on emission. This is the energy that is aligned with the mid-point of the two experimental hyperfine energies for a given level by conducting an automatic search for the appropriate *Kalpha*, the fine-structure constant multiplier.

The Doppler shift is applied to all states, but although all emissions are to occur from the same coordinates on the $1S_{1/2}$ ring to which all levels collapse, and use the same angle $\theta$, see Figure 8.3, this shift is not the same for all states. The reason for this is that the proton speed that enters into the Doppler shift is determined by the energy level, which in turn determines the field affecting the proton and its linear velocity. This energy varies for each $(n_\phi, n_r)$ combination, and also with the value of *Kalpha*.[21]

The main part of the program is given in the *Kloop* search for *Kalpha* in lines 300–348. The aim is to align the MVR fine-structure frequency with the mid-point of the two pre-Doppler emission experimental hyperfine frequencies. This is carried out by varying the fine-structure constant with a decimal-place search for its multiplier *Kalpha*. There are two passes through the *Kloop* corresponding to two different *Kalpha* searches as follows.

*First pass.* This is centered on the ground state to which all levels collapse for energy transactions. At the trial value of *Kalpha* in the decimal-place search, the coordinates are calculated at a point on the circumference of the $1S_{1/2}$ ring when the circular ring is at the boundary of oscillation, subject to the condition 'field energy equals electron spin-3 energy' (which is analogous to 'P.E. equals K.E.') in line 627. In the program, these coordinates are $x/\bar{r}_{gs} =$ *Xstore* and $y/\bar{r}_{gs} = gamma$. They will be crucial in determining the emission angle in the Doppler calculation. The value of *Kalpha* that results from the decimal-place search is determined from the condition that the difference between the MVR fine-structure frequency *VortexEnergy*1 and the mid-point of the two experimental hyperfine frequencies for the ground state is to be minimized. This appears in line 319, *after* the Doppler calculation, so that there can be a correcting division by *Doppler*1 from line 704 or 706, to arrive at a pre-Doppler value. We should remind ourselves that the Doppler shift only occurs on emission and the fine-structure calculation is pre-emission.

---

[21] Although two states share the same elliptical ring, the requirement that the fine-structure energy for each must be aligned with the hyperfine pair mid-point for that state gives two different values for *Kalpha*. The problem is to justify these two different values and the conjecture advanced here is that for an elliptic ring one emission is from the perigee and the other from the apogee.

In summary, on emerging from the first pass of *Kloop*, we should have the coordinates of the emission point on the ground state ring, output (6)–(8), which depend on the *Kalpha* under search. These supply the Doppler calculation so that we then know what blue-shift multiplier to apply to the fine-structure frequency.

*Second pass.* The focus here is on the actual level that is input and the search is now for a different *Kalpha*. This value is subject to the condition that the difference between the MVR fine-structure frequency *VortexEnergy1* and the mid-point of the two experimental hyperfine frequencies for the *chosen* input level is to be minimized. This appears in line 321. Again, the two experimental hyperfine values are pre-Doppler corrected, again using the ground state value *Doppler1*, since all levels collapse to the ground state for emission.[22] What emerges from the second pass is a value of *Kalpha*, output item (24), that obtains this hyperfine mid-point alignment for the chosen state. The pre-Doppler MVR fine-structure value, item (12), should then be in agreement with the mid-point of the pre-Doppler hyperfine values for the input level, item (13).

### 8.3.2   *Data*

The data for selected hydrogen atom level frequencies (MHz) have been taken from the work of Horbatsch and Hessels (2016) who have tabulated pairs of hyperfine energy values (MHz) for $nS_{1/2}$, $nP_{1/2}$, $nP_{3/2}$, $nD_{3/2}$, and $nD_{5/2}$ states. These appear in lines 10–155 up to $n = 6$. The physical constants in lines 238–245 have been taken from the CODATA (2018) values as published by Mohr *et al.* (2018).[23]

---

[22]In other words, we assume that the experimental hyperfine values have undergone a Doppler shift and so we reverse its effect to get their hypothetical pre-emission frequencies. The MVR fine-structure value will be multiplied by the Doppler factor, and the hyperfine frequencies will be restored to their experimental values when the emission frequencies are output.

[23]For a full list of CODATA values, see physics.nist.gov/constants.

### 8.3.3 *Inputs*

(1) This is for the state code. For example, the $3S_{1/2}$ state is to be entered as '3$S$1/2' (omitting the quotation marks) when prompted by the program run followed by <return>.

(2) The '+' or '−' refers to the rotation direction of the electron's spin-2 circuit. For the former, the proton and electron spin-2 field momenta rotate in the same sense, whereas for the latter they are opposite. The former gives the lower of the two hyperfine values, while the latter gives the higher one.

(3) The third input is the counterclockwise rotation angle from the positive $x$ axis of the electron spin-2 ellipse major axis. Here, the origin is taken as the leftmost of the two foci. An angle of 135 degrees yields successful searches for most levels.

(4) This input selects the type of Doppler shift on emission, the options being 'red' or 'blue'. The MVR fine-structure value is still aligned with the average value of the two hyperfine frequencies for that level after a *Kalpha* search. However, the red shift results in *Kalpha* > 1 and the blue shift has *Kalpha* < 1.

### 8.3.4 *Automatic input check*

The input checks occur in lines 198–220. For the state code, if $n$ is not changed from the initial $n = 0$ by the selection of one of the cases in the program, then the input is void. The input angle $\tau$ must satisfy $0 \leq \tau \leq 360°$. The only choices available for the second and fourth inputs are clearly stated when the program is run.

### 8.3.5 *Sommerfeld–Dirac fine structure*

The calculation is based on the CODATA fine-structure constant $\alpha$, so *Kalpha* = 1. Given the Sommerfeld quantum numbers, $n_\phi = L + Sp + 1/2$ and $n_r = n - n_\phi$, then the calculation for the positive emission fine-structure frequency *DiracEnergy* (MHz) for a state is called by the subroutine [*Sommerfeldfs*] in line 245. The calculation proceeds in lines 570–580 as follows:

$$SommEnergy = \frac{m_o c^2}{\left(1 + \frac{m_o}{m'_o}\right)} \left(1 - \frac{1}{\left(1 + \frac{\alpha^2}{Y^2}\right)^{1/2}}\right) \frac{10^{-6}}{h} \quad (8.10)$$

This is expressed in line 579 in terms of the Rydberg frequency as

$$SommEnergy = \frac{2 \times Rydfreq}{\alpha^2 \left(1 + \frac{m_o}{m'_o}\right)} \left(1 - \frac{1}{\left(1 + \frac{\alpha^2}{Y^2}\right)^{1/2}}\right) \quad (8.11)$$

where $m_o$ and $m'_o$ are the electron and proton rest masses, and

$$Rydfreq = \frac{m_o \alpha^2 c^2 \times 10^{-6}}{2h} \quad (8.12)$$

is given in line 240. The computer output appears as (1), see Appendix D.2.

### 8.3.6   *Mass vortex ring fine structure*

In the MVR calculation, the possibility is introduced of varying the fine-structure constant $\alpha$ through *Kalpha*, with the adjusted fine-structure constant being given by $\alpha_1 = Kalpha \times \alpha$. With this in mind, we define $Y_1 = n_r + (n_\phi^2 - \alpha_1^2)^{1/2}$. The computation for the positive emission frequency MVR in the subroutine [*MVRfs*] from lines 581–590 is as follows:

$$MVR = \frac{m_o \alpha_1^2 c^2 10^{-6}}{2h \left(1 + \frac{m}{m'_o}\right) (\alpha_1^2 + Y_1^2) \times vortex5} \quad (8.13)$$

where the relativistic electron mass $m$ in the reduced mass term modifies the electron rest mass $m_o$ according to

$$m = \frac{m_o}{vortex5} \quad (8.14)$$

Here, *vortex*5 is given in line 585 by

$$vortex5 = \left(1 - \frac{\alpha_1^2}{2Y_1^2 \left(1 + \frac{\alpha_1^2}{Y_1^2}\right)}\right)^{1/2} \tag{8.15}$$

Line 589 in the computer program expresses (8.13) in terms of the Rydberg frequency (8.12) as follows:

$$MVR = \frac{Rydfreq \times (Kalpha)^2}{\left(1 + \frac{m}{m_o'}\right)(\alpha_1^2 + Y_1^2) \times vortex5} \tag{8.16}$$

This is *erad* in the program. We should bear in mind that the Rydberg frequency contains the square of the unadjusted fine-structure constant and so needs the square of *Kalpha* in (8.16) for the adjusted version. The computer output for *Kalpha* = 1 appears as output (2), see Appendix D.2.

### 8.3.7 *Emission coordinates*

We posit here that although states with various $(n_\phi, n_r)$ adopt either an elliptic or circular OAM ring configuration, and oscillate over the much smaller proton ring in that configuration, for an emission or absorption to occur, a state must temporarily collapse to the circular $1S_{1/2}$ ring. So, the $(x/\bar{r}_{gs}, y/\bar{r}_{gs})$ coordinates at the point of emission on the circumference of this circular ring are crucial to the calculation of the radiation Doppler shift. These are expressed as multipliers of the electron ground state radius $\bar{r}_{gs}$, are represented in the computer program as $(Xstore, gamma)$, and appear in the subroutine [*coordinates*] in lines 600–642. For a Doppler blue shift, the radiation proceeds from its emission point away from the proton along the produced line joining the spin-2 centers of the proton and electron; see Figure 8.3. As the radiation emerges from the atomic system, the proton is advancing toward it. For a red shift, the ray passes through the proton. So the angle of emission with respect to the proton velocity vector must be taken into account.

For a given state, the [*coordinates*] subroutine is also used to find the coordinates of the perigee and apogee, or circle circumference, at the extreme end of its oscillation. Since these are not crucial to the energy calculation, the fine-structure constant is not adjusted and *Kalpha* = 1. We should also point out that the condition for locating the *Xstore* coordinate (proton–electron separation) — that the electron ring energy equals the proton field energy in the ring plane — has been calculated using a Coulomb law for circular states only. Lines 349–377 handle this procedure and the computer output appears as items (3)–(5), for which there are two sets of values for the non-circular ring possessing a perigee and apogee.[24]

The degeneracy of fine-structure levels is not resolved in MVR theory by a mechanism known as 'spin' (with possible values $\pm 1/2$) as in quantum mechanics. They are distinguished by two different values of *Kalpha*.[25] Nevertheless, the quantum mechanics numbers can be used to classify the ellipses. If the spin quantum number $Sp = +1/2$ and the orbital quantum number $L = n - 1$, then the ring is circular, that is, when $n = n_\phi$. If $L \neq n - 1$, then the ring is elliptical. So, the first task of the computer program is to establish the ring type.

### 8.3.7.1   *Calculation of the $\Gamma$ coordinate*

The $\Gamma = y/\bar{r}_{gs}$ or *gamma* coordinate is the ring radius expressed as a multiplier of the ground state radius. The calculation (7.22) is independent of the phase of the bound-state oscillation, and is represented in the [*coordinates*] and [*printcoords*] subroutines. In lines 605–607 of the program, the radius *gamma* in Table 8.2 is given as a fraction of the $1S_{1/2}$ radius $\bar{r}_{gs}$ — obtained from (7.23) at $n_\phi = 1$ — using the equation of an ellipse

$$gamma = \frac{a(1 - \epsilon^2)}{(1 + \epsilon \cos \theta)\bar{r}_{gs}} = \frac{n_\phi^2 \left(1 - \frac{\alpha_1^2}{n_\phi^2}\right)}{(1 + \epsilon \cos \theta)(1 - \alpha_1^2)} \qquad (8.17)$$

---

[24] In item (5), the square of the spin-2 radius from the proton $Y$ includes a subtraction of the proton spin-3 radius.

[25] These could be related to the apogee and perigee radii, but no investigation has been carried out to support that conjecture.

Table 8.2   Selection of states and their conjectured association with a circular ring, or the perigee/apogee of an elliptical ring with *Kalpha* = 1.

| Ring | $n_\phi$ | $n_r$ | *Xstore* ($=x/\bar{r}_{gs}$) | $\Gamma(=y/\bar{r}_{gs})$ | c/p/a |
|------|------|------|------|------|------|
| $1S_{1/2}$ | 1 | 0 | 1.732 412 679 5 | 1.0 | circle |
| $2S_{1/2}/2P_{1/2}$ | 1 | 1 | 7.982 170 656 9 | 0.535 897 281 2 | perigee |
| | 1 | 1 | 2.879 901 407 4 | 7.464 315 734 2 | apogee |
| $2P_{3/2}$ | 2 | 0 | 6.929 036 835 7 | 4.000159 762 6 | circle |
| $3S_{1/2}/3P_{1/2}$ | 1 | 2 | 17.992 919 665 9 | 0.514 718 071 5 | perigee |
| | 1 | 2 | 4.274 484 432 8 | 17.485 920 975 9 | apogee |
| $3P_{3/2}/3D_{3/2}$ | 2 | 1 | 17.854 717 094 6 | 2.291 885 866 0 | perigee |
| | 2 | 1 | 8.791 380 891 1 | 15.708 912 946 7 | apogee |
| $3D_{5/2}$ | 3 | 0 | 15.590 061 493 1 | 9.000 426 033 5 | circle |
| $4S_{1/2}/4P_{1/2}$ | 1 | 3 | 31.996 506 856 5 | 0.508 066 282 4 | perigee |
| | 1 | 3 | 5.678 557 653 2 | 31.493 211 813 9 | apogee |
| $4P_{3/2}/4D_{3/2}$ | 2 | 2 | 31.930 180 792 6 | 2.143 678 052 2 | perigee |
| | 2 | 2 | 11.518 378 694 1 | 29.857 813 064 8 | apogee |
| $4D_{5/2}/4F_{5/2}$ | 3 | 1 | 31.540 584 533 | 5.417 249 129 5 | perigee |
| | 3 | 1 | 17.817 195 552 2 | 26.584 312 993 3 | apogee |
| $5S_{1/2}/5P_{1/2}$ | 1 | 4 | 49.998 358 972 8 | 0.505 102 350 3 | perigee |
| | 1 | 4 | 7.084 398 814 2 | 49.497 027 811 6 | apogee |

*Note*: The proton–electron displacement *Xstore* and the electron OAM ring radius $\Gamma$ are given as a fraction of the electron $1S_{1/2}$ circular ring radius $\bar{r}_{gs}$. These are the computer program outputs (3) and (4), and are independent of inputs (2), (3), and (4).

Here, $a$ is given by (7.24), the eccentricity $\epsilon$ by (7.25), the unadjusted fine-structure constant $\alpha_1 = \alpha \times Kalpha$ (*Kalpha* = 1), where $\theta = 0$ corresponds to the perigee and $\theta = \pi$ to the apogee in line 605.

The calculated values are presented in Table 8.2 with *Kalpha* = 1 (no adjustment to the fine-structure constant). We can see from this table which levels share the same elliptical ring.

### 8.3.7.2   *Calculation of the Xstore coordinate*

The major part of the [*coordinates*] subroutine determines the $Xstore = x/\bar{r}_{gs}$ coordinate subject to the condition (7.56) as represented in line 627 of the program. This ensures that the proton spin-3 field energy in the electron ring plane is equal to the electron energy remaining in the ring. It is analogous to, but not identical to, the condition that potential energy equals kinetic energy.[26] The program conducts an automatic decimal-place search designed to minimize *Fsearch* in line 627. This is achieved by a variation of $X = Xstore = x/\bar{r}_{gs}$ in (8.20) below (line 613 of the program) in which only *Fint* from (8.20)–(8.22) is a function of $X$:

$$Fsearch = \frac{FPX}{(1 + M \cdot FX \cdot FPX)} - \frac{1}{2(Y_1^2 + \alpha_1^2)} \qquad (8.18)$$

This equation is a representation of (7.56) in which $M$ is the ratio of electron rest mass. Based on (7.43), we have

$$FX = \frac{\left(1 - \frac{\alpha_1^2}{2n_\phi^2}\right)^{1/2} \times Fint}{Y_1 \left(1 - \alpha_1^2\right)^{3/2} \times gamma} \qquad (8.19)$$

where $\alpha_1^2 \to \alpha_1^2/2$ at the oscillation boundary.

The integral calculation *Fint* is taken from (7.55) with the $2\pi$ and $\bar{r}_3$ omitted, having been absorbed into the $FP$ and $FPX$ calculations. This is set up in program lines 620–625 using (7.54) as follows:

$$XX = \frac{X}{gamma}$$

$$R1 = \frac{M}{gamma} \qquad (8.20)$$

$$R2 = \frac{\alpha_1}{gamma \times (1 - \alpha_1^2)}$$

Then, if we define $A$ and $B$ as follows,

$$A = \frac{1}{(1 - R1) \left(1 + \frac{XX^2 + R2^2}{(1-R1)^2}\right)^{1/2}}$$

$$B = R2^2(XX^2 + (1 - R1)^2) \qquad (8.21)$$

---

[26]It is not identical because part of the mass of the electron in Equation (6.18) is provided by its self-potential or spin-3 curvature. As the spin-3 radius becomes arbitrarily large, the mass tends to its rest mass, and the SAM linear trajectory becomes a pure *optical* OAM.

the integral can be approximated to

$$Fint = A\left(1 + \frac{3}{4}A^4B + \frac{105}{64}A^8B^2\right) \qquad (8.22)$$

This calculation can be found in lines 620–625 of the program. The crucial coordinates calculation is for the ground state, in which case, the subroutine [*coordinates*] called from line 313 is included in a decimal-place search for *Kalpha* and $\alpha_1 = \alpha \times Kalpha$ (*Kalpha* < 1).

### 8.3.8 *The Doppler shift multiplier*

Section 8.2.4 illustrates the concept of the Doppler shift, in which the radiation emission from a point on the electron ground state OAM ring moves either towards or away from the proton along the direction of the line joining the proton and electron spin-2 axes. The angle this line of emission makes with the proton–electron OAM ring axes (program line 702) is crucial for determining the component of the proton motion along the emission line. The electron ring is stationary in the center of mass frame. If the emission radiates away from the proton which is advancing toward it, then there is a blue shift. However, if the emission runs through the proton, when it emerges on the other side, the proton is receding from it so there is a red shift.

So, the idea is to calculate the proton velocity at the entrance to the bound state, which through momentum conservation depends on the electron velocity, and then take the component of the radiation emission relative to the proton. Momentum conservation requires that $mv_x = m'v'_x$, where $v'_x$ is the required proton linear speed in the center of mass frame, and $M = m_o/m'_o$ is the electron to proton rest mass ratio.[27] The electron linear speed ratio $v_x/c$ is the first factor of the right-hand side of (8.23). The mass $m_o$ of the electron has

---

[27]The linear speed of the electron at the oscillation boundary is $v_x = 0$, while the proton speed is $v'_x > 0$. This appears to be a violation of the momentum conservation law, but their oscillations are $\pi/2$ out of phase. This means that when the proton speed is zero, the electron speed is at its maximum.

been made relativistic in the second quotient but not the mass of the proton.

From lines 700–701, we have

$$\frac{v'_x}{c} = dopplera \times dopplerb = \frac{\alpha_1}{2(Y_1^2 + \alpha_1^2)^{1/2}} \frac{M}{\left(1 - \frac{\alpha_1^2}{2(Y_1^2 + \alpha_1^2)}\right)^{1/2}}$$

(8.23)

We then take the proton velocity component along the radiation direction; thus

$$\frac{v'_x}{c} \cos\theta = dopplera \times dopplerb \times \cos\theta$$

(8.24)

so that using (8.17), in line 702, we have

$$\cos\theta = \frac{Xstore}{(Xstore^2 + (gamma - M)^2)^{1/2}}$$

(8.25)

We note that (A.15) provides the justification for the $M$ in the denominator of (8.25) in which the proton spin-3 radius is subtracted from the electron spin-3 radius, in order to have the proton spin-2 axis as the origin. The relativistic Doppler factor given on lines 704 (blue shift) and 706 (red shift) is now

$$Doppler1 = \frac{\left(1 \pm \frac{v'_x}{c}\cos\theta\right)}{\left(1 - \left(\frac{v'_x}{c}\cos\theta\right)^2\right)^{1/2}}$$

(8.26)

The complete Doppler blue-shifted (+) or red-shifted (−) fine-structure frequency then appears in line 818 of the [*printout1*] subroutine as *VortexEnergy1* × *Doppler1*. This has been aligned, through a decimal-place search of *Kalpha* to modify the fine-structure constant, with the mid-point of the two experimental hyperfine frequencies for the level under consideration. All that remains is to account for the positive/negative deviation between this mid-point and the upper/lower hyperfine frequency. This is the task of the next chapter. Meanwhile, we now investigate the bound-state oscillations.

## 8.4 Electron States

### 8.4.1 *The oscillations*

We now treat the mechanism of the bound state oscillation. Here, we shall see that the proton and electron oscillations share the same time period, but due to the changes in the field inhabiting one ring from the other, one oscillation lags $\pi/2$ in phase behind the other. So, while there is a conservation of total energy (linear and 'potential') in each ring, the conservation of linear momentum breaks down inside the bound state, that is, in the oscillation region. In Section 6.4.2.1, it was suggested that outside the oscillation boundary, the spin-2 axis on the helical trajectory remains aligned with the ring plane. This preserved the conservation of energy law. It is now posited that inside the boundary, the spin-2 axis is aligned with the trajectory. So the field effect of the host diminishes as the linear momentum increases. Since all transitions occur at the boundary, then the reduced mass which depends on linear momentum conservation is assumed to hold. The electron's emission at the boundary changes its linear momentum and brings it to rest in the center of mass frame. However, the electron's emission at the boundary has not changed the proton's own momentum in the center of mass frame, so it behaves as if it is still in a linear momentum conservation relationship with the electron. According to this argument, the reduced mass effect, which is calculated with respect to the proton frame, should still apply at the boundary after the electron's emission as it enters the bound state.

### 8.4.1.1 *Stage (a)*

In Figure 8.4(a), we visit the participants in the center of mass frame at the moment when the electron, in its $1S_{1/2}$ state, has reached the entrance to the bound state at a distance of $\sim\sqrt{3}\bar{r}_3$ from the proton. On entering the binding, the electron radiates away its linear energy given by the Doppler-shifted and reduced-mass adjusted (7.17), with $n_\phi = 1$, and $n_r = 0$. The electron is brought to rest, $v_x = 0$, and it is left with an energy in its ring calculated as follows. We take the field-free electron ring energy given by (7.15) which applies as

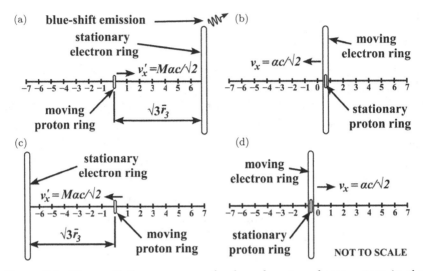

Figure 8.4   The oscillation sequence of a bound proton–electron state in the hydrogen ground state. (a) The electron, which approaches the smaller proton ring from the right, radiates and comes to rest, while the proton still moves toward it with speed $v'_x = Mac/\sqrt{2}$. (b) The proton's movement toward the electron raises the proton field in the latter, and causes the electron to move toward the proton until the rings are concentric (proton inside electron). The electron spin-3 momentum is completely transferred into linear momentum, and with no electron field originating from its ring plane to affect the proton, the latter comes to rest. The electron moves at speed $v_x = \alpha c/\sqrt{2}$. (c) The electron's movement away from the proton reduces the proton field in the electron, converts electron linear energy of motion back into ring energy, and slows it to rest. This increases the electron spin-3 momentum as well as its generated field, and consequently the electron field in the proton, so that the latter moves toward the electron. (d) This is the mirror image scenario of (b).

$x \to \infty$ and subtract the proton field energy (7.38) inhabiting the ring plane at $x \sim \sqrt{3}\bar{r}_3$, with $n'_\phi = n_\phi = 1$ and $n'_r = 0$, so that to order $\alpha^4$ we have the electron ring energy as

$$\mathcal{E} = m\alpha^2 c^2 - \frac{m\alpha^2 c^2}{2} = \frac{m\alpha^2 c^2}{2} \tag{8.27}$$

and its energy of motion is zero.[28] One half of the electron ring energy has now been radiated away, with Doppler and reduced mass

---

[28] Note the electron ring energy is $\varepsilon = p(\alpha c/\sqrt{2}) = (m\alpha c/\sqrt{2})(\alpha c/\sqrt{2}) = m\alpha^2 c^2/2$.

adjustments included to refer to the proton rest frame. Equation (8.27) is then the electron's total energy to be conserved during the oscillation. During the oscillation, it will be distributed between resident ring energy and linear energy of motion, the latter being displaced from the electron ring and equaling the energy of the proton field inhabiting the electron ring.

The proton, which has not radiated, is still moving at a speed $v'_x = Mac/\sqrt{2}$, calculated from the conservation of linear momentum at the entrance to the bound state just prior to the electron radiating. So, already we see a phase difference between the linear-speed oscillations in that the electron is stationary at the oscillation boundary and the proton is in motion toward it.

At the boundary, we know from (6.30) and (6.35) that the electron spin-3 field energy in the proton ring and the proton spin-3 field energy in the electron ring are equal. So, the electron field energy in the proton ring is

$$E_{3f} = \frac{m\alpha^2 c^2}{2} \tag{8.28}$$

This energy displaces the equivalent energy from the proton ring into the motion of the proton.

The proton energy in its own ring is calculated as follows. We take the energy given by (7.15) with the proton mass $m'$ as $x \to \infty$, and subtract the electron field energy (8.28) at $x = \sqrt{3}\bar{r}_3$, with $n_\phi = 1$, so that to order $\alpha^4$ we have the proton ring energy as

$$\mathcal{E}' = m'\alpha^2 c^2 - \frac{m\alpha^2 c^2}{2} \tag{8.29}$$

During the proton ring oscillation, as the proton–electron displacement decreases, the proton field energy in the electron ring increases, and the electron's resident ring energy is displaced into linear energy. Consequently, the electron's resident ring energy from which its field arises diminishes. We recall that this is due to the spin-2 axis aligning itself with the helical trajectory and diminishing its field-generating effect. So, the proton ring receives a decreasing electron field effect that eventually diminishes to zero near the proton. This means that the proton's resident ring energy will vary between

$$m'\alpha^2 c^2 - \frac{m\alpha^2 c^2}{2} \leq \mathcal{E}' \leq m'\alpha^2 c^2 \qquad (8.30)$$

An illustration of this first stage is shown in Figure 8.4(a).

### 8.4.1.2 *Stage (b)*

Figure 8.4(b) illustrates the second stage of the oscillation. Since the proton is moving toward the stationary electron, the proton field energy in the electron is increasing, which causes the latter to move toward the proton. They meet with their rings concentric. At this point, the electron is at its maximum speed, and all its resident ring energy has been converted into linear energy of motion. Consequently, it cannot generate a field to affect the proton, and so the proton has no displacement of its ring energy into motion energy and it comes to rest with $v'_x = 0$.

In the center of mass frame, the proton field energy in the electron ring is

$$\varepsilon'_{3f} = m\alpha^2 c^2 \qquad (8.31)$$

Here, the electron ring is entirely occupied by proton field energy, and the remaining electron ring energy has been displaced entirely into energy of motion

$$\mathcal{E} = \frac{m\alpha^2 c^2}{2} \qquad (8.32)$$

### 8.4.1.3 *Stage (c)*

With the electron passing over and moving away from the proton, there is a gradual reduction of proton field energy in the electron ring. The electron's linear energy of motion is therefore returned into resident electron ring energy and the electron slows. As the electron ring energy accumulates, and its spin-2 axis becomes more aligned with its ring plane, its field effect on the proton increases and so it attracts the proton toward it. So, the proton begins to follow the electron, albeit at a much slower speed. This is shown in Figure 8.4(c). Eventually, all the electron's linear energy of motion is returned to its ring and it stops at the oscillation boundary when $|x| = \sqrt{3}\bar{r}_3$. At this point, its resident ring energy is again (8.28) and the proton is moving with speed $v'_x = M\alpha c/\sqrt{2}$ toward it. We now have the mirror image of Stage (a).

### 8.4.1.4 *Stage (d)*

This is depicted in Figure 8.4(d) as the mirror of Stage (b). With the electron stationary, the proton continues to move toward the electron, raising the proton field energy in the electron ring. This displaces electron ring energy into linear energy of motion and causes the electron to move toward the proton. The increase in electron linear energy reduces its field effect on the proton until the rings are concentric again, at which point all the electron's ring energy has been converted into linear motion. With no electron field to affect it, the proton stops.

It should be clear from this scenario that the proton oscillation velocity lags the electron oscillation velocity by $\pi/2$.

## 8.4.2 *The OAM mass rings*

For the MVR model of the hydrogen atom, there are pairs of elliptical rings, one member of the pair for each state, that share the same $(n_\phi, n_r)$ values; see Table 8.2. These are shown in Figure 8.5. Is there only one elliptic ring for the two states (e.g., $2S_{1/2}$ and $2P_{1/2}$) with two modes of emission presenting different fine-structure energies, or are there two separate but otherwise identical rings? What is it that distinguishes the two members of the pair, one from the other? In quantum mechanics, this question was answered by the introduction of the spin quantum number $Sp$. So, for example, for the pair $(n_\phi, n_r) = (1, 1)$, the $2S_{1/2}$ state arises from $n_\phi = l + Sp + 1/2$, with $l = 0$ and $Sp = 1/2$, and $2P_{1/2}$ from $l = 1$ and $Sp = -1/2$.

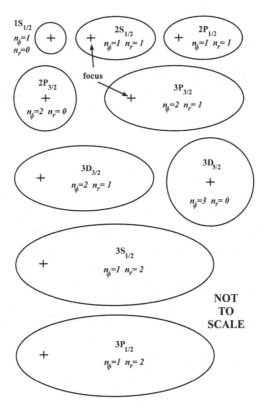

Figure 8.5   The OAM mass rings associated with the first nine fine-structure states, from Table 8.2.

Since there is no spin quantum number in MVR theory, there must be another explanation for the two different energies: the hyperfine average for the $2S_{1/2}$ state is at $822\,025\,488.313\,8$ MHz, and the $2P_{1/2}$ state is $822\,026\,516.551\,2$ MHz.

A suggestion that can be made here, and it seems the most economical one, is that there is only one elliptic ring to a pair of states, and the SAM mass cluster emits from the perigee for one state and the apogee for the other. For example, for $(n_\phi, n_r) = (1, 1)$, since the smaller the radius, the larger the frequency, $2S_{1/2}$ must be the apogee and the $2P_{1/2}$ the perigee; see Table 8.2.[29] The two

---

[29]After all, the smallest radius is at the ground state and has a frequency of $3\,288\,087\,212.213\,1$ MHz, the highest frequency of the set of states.

hyperfine values for a state are then related to the two possible spin-2 senses $\pm 1\hbar$ for each state; see Chapter 9. This gives four possibilities for an elliptic ring. For a circular ring such as $2P_{3/2}$, we have one emission radius with two possible spin-2 rotation senses, so only two possibilities.

If this conjecture has weight, then there might be some relation between the emission radius of the ellipse (at the perigee or apogee) and the value of *Kalpha* which multiplies the fine-structure constant upon which $v_3$ depends. For $(n_\phi, n_r) = (1, 1)$, the perigee is at $0.535\,897\,281\,2$ and the apogee at $7.464\,315\,734\,2$ (multiplied by the ground state radius); see Table 8.2. For a red shift, for $2S_{1/2}$, we have *Kalpha* $= 1.000\,000\,001\,744$ and for $2P_{1/2}$, it is *Kalpha* $= 1.000\,000\,627\,163$. For a blue shift, for $2S_{1/2}$, we have *Kalpha* $= 0.999\,998\,784\,660$ and for $2P_{1/2}$ *Kalpha* $= 0.999\,999\,410\,077$. In both cases, for *Kalpha* $> 1$ (red shift) and *Kalpha* $< 1$ (blue shift), the greater speed is at the perigee for $2P_{1/2}$ as one might expect. So, both a red shift and a blue shift are possible though not simultaneously so.

Of course, a model could be employed in which the Doppler effect is abandoned altogether. Here, the emission must then take place perpendicular to the proton–electron axis. There would then be no need for higher levels to collapse to the ground state ring for transitions. We would then need a different set of *Kalpha* values. An investigation into how these values arise, in either model, promises to be a substantial research project on its own. So, it is better to stop at this vantage point where the terrain is reasonably clear, and let others digest its possibilities, than march forward with abandon into a new unchartered wilderness.

## 8.5   Electron Configuration

### 8.5.1   *Quantum mechanics*

In 1919, Irving Langmuir remarked that

> In attempting to determine the arrangement of electrons in atoms we must be guided by the numbers of atoms which make up the inert gases; in other words by the atomic numbers of these elements,

namely, helium 2, neon 10, argon 18, krypton 36, xenon 54, and niton 86 (1919, 869).

His scheme distributes the electrons on shells having equal thickness, a set of concentric spheres, with radii in the ratio $1 : 2 : 3 : 4 \ldots$. The surface areas of the shells are consequently in the ratio $1^2 : 2^2 : 3^2 : 4^4 \ldots$ which is the ratio $2 : 8 : 18 : 32 \ldots$ of the number of cells that he assigns to each shell. The innermost shell can contain one electron to each cell. This gives the two electrons for helium. For all other outer shells, each cell on the shell has first to be given one electron each, then each cell receives a second electron until all cells on that shell have two each. So, when in addition to the two electrons on the innermost shell, the eight cells on the second shell are given one electron each, we arrive at the 10 electrons for neon. Completing the second shell with another eight electrons, so that each cell now has two electrons each, gives the 18 electrons for argon. The next assignment is 18 electrons to the third shell, one to each cell, to total the 36 electrons of krypton. Doubling the electrons in the third shell with another 18 produces the 54 electrons for xenon. To distinguish between the two electrons in a single cell, Langmuir gives each shell two layers, an inner and outer one, so that the two electrons have different distances from the nucleus (1919, 871). However, rather like Thomson's (1904) plum pudding model, the scheme suffers from the defect that the electrons are unrealistically at rest. Also, as Bohr (1921) later pointed out, the assumption of polyhedral symmetry of the cells rested on the assumption that the electrons on a shell obtained a simultaneous binding rather than successive one.

A modification of this scheme, devised to vitalize the electrons with motion, was published by Bohr (1923). Sommerfeld's principal and azimuthal quantum numbers, $n$ and $k$, are adopted so that an orbital trajectory is denoted by $n_k$. It is not possible to have $k = 0$ for then the orbit is a straight line that passes through the nucleus. Neither can the azimuthal rotation component $k$ exceed $n$. The possible orbits are then $1_1, 2_1, 2_2, 3_1, 3_2, 3_3, \ldots$ which are filled with electrons to correspond with the elements in the periodic table. For example, hydrogen has one electron in $1_1$, while carbon has two in $1_1$, three in $2_1$, and one in $2_2$. The difficulty with Bohr's scheme is that while $1_1$ is confined to a maximum of two electrons, both $2_1$ and $2_2$ can take four each, and $3_1$ can take six (Bohr 1922). Without any geometrical justification for these maximum values, the choice seems rather arbitrary.

Sommerfeld does not entertain a detailed theory as to how the electrons are configured into orbits. He merely suggests that

> the periodic numbers $2, 8, 8, 18, 18, 32$ give us an index of the probable numbers of the electrons that occupy the successive inner shells. (Sommerfeld 1923, 108)

This allows the last entry in each period helium, neon, argon, krypton, xenon, and radon to take on the successive sums of these values $2, 10, 18, 36, 54, 86$.

In a quantum mechanical system, the labeling of the sub-shells is led by the terms 'sharp' $l = 0$, 'principal' $l = 1$, 'diffuse' $l = 2$, and 'fundamental' $l = 3$, where $l$ is the orbital quantum number. Here, the maximum number of electrons follow the rule $2(2l+1)$. The factor of 2 counters degeneracy by positing two different electron spins $\pm 1/2$. The notation employs a superscript to indicate the number of resident electrons. By way of illustration, hydrogen ($Z = 1$) is $1s^1$ and lithium ($Z = 3$) is $1s^2 2s^1$. The electron configurations of the noble gases are as follows:

He: $1s^2$
Ne: $1s^2 2s^2 2p^6$
Ar: $1s^2 2s^2 2p^6 3s^2 3p^6$
Kr: $1s^2 2s^2 2p^6 3s^2 3p^6 3d^{10} 4s^2 4p^6$
Xe: $1s^2 2s^2 2p^6 3s^2 3p^6 3d^{10} 4s^2 4p^6 4d^{10} 5s^2 5p^6$
Rn: $1s^2 2s^2 2p^6 3s^2 3p^6 3d^{10} 4s^2 4p^6 4d^{10} 5s^2 5p^6 4f^{14} 5d^{10} 6s^2 6p^6$

These examples give an idea of the order in which the shells are filled with electrons.

Table 8.3 is reproduced from Johnson (1952, 95) and shows that in the quantum mechanical scheme, the electron pair is denoted by $m_s = \pm 1/2$. For example, for $n = 2$, we have $m_l = 0$ at $l = 0$ and

$m_l = \pm 1, 0$ at $l = 1$ which makes four possibilities, which is doubled to eight electrons with the two values of $m_s$ in each case.

### 8.5.2   *MVR theory*

The quantum mechanical scheme conveyed in Table 8.3 can be reorganized in terms of the numbers $(n_\phi, n_r)$ if each pair of numbers is assigned two electrons. These can be distinguished by their spin-2 rotation senses $\pm 1\hbar$. A circular ring would have only one orientation and account for two electrons. An elliptic ring specified by $(n_\phi, n_r)$ would have two electrons with opposite spin-2 rotations and $2n_\phi + 1$ orientations specified by $m_\phi$. For example, for $n_\phi = 2$ , we could have $m_\phi = \pm 2, \pm 1, 0$. This scheme is set out in Table 8.4.

As we move up the atomic numbers, the circular ring at a given $n$ is filled first, and then the other levels are occupied in order of ascending $n_\phi$. So, for each elliptic-ring pair $(n_\phi, n_r)$, there are two

Table 8.3   The $m_l$ values for electron configurations denoted by four quantum numbers $n, l, m_s, m_l$ (Johnson 1952, 95).

| $n$ | $m_s$ | $l = 0$ | $l = 1$ | $l = 2$ | $l = 3$ | Total of electrons in shell |
|---|---|---|---|---|---|---|
| 1 | $\pm 1/2$ | 0 | | | | 2 |
| 2 | $\pm 1/2$ | 0 | $\pm 1, 0$ | | | 8 |
| 3 | $\pm 1/2$ | 0 | $\pm 1, 0$ | $\pm 2, \pm 1, 0$ | | 18 |
| 4 | $\pm 1/2$ | 0 | $\pm 1, 0$ | $\pm 2, \pm 1, 0$ | $\pm 3, \pm 2, \pm 1, 0$ | 32 |

Table 8.4   The $m_\phi$ values for electron configurations denoted by two quantum numbers $(n_\phi, n_r)$ and the two spin-2 rotation senses.

| $n$ | spin-2 | $n_\phi = 1$ | $n_\phi = 2$ | $n_\phi = 3$ | $n_\phi = 4$ | Total of electrons in shell |
|---|---|---|---|---|---|---|
| 1 | $\pm 1$ | 1 | | | | 2 |
| 2 | $\pm 1$ | $\pm 1, 0$ | 1 | | | 8 |
| 3 | $\pm 1$ | $\pm 1, 0$ | $\pm 2, \pm 1, 0$ | 1 | | 18 |
| 4 | $\pm 1$ | $\pm 1, 0$ | $\pm 2, \pm 1, 0$ | $\pm 3, \pm 2, \pm 1, 0$ | 1 | 32 |

electrons. From Table 8.4, it seems sufficient that there is only one elliptic ring for an elliptic-ring pair $(n_\phi, n_r)$, and that it is occupied by 2 electrons which are distinguished by their spin-2 rotation senses.

## 8.6 Nuclear Configuration

There is an opportunity here to account for the prevalence of alpha particles in radioactive emissions. The periodic table can be adequately accounted for in MVR theory by positing two electrons to a mass ring with opposing spin-2 rotation senses; see Section 8.5.2. However, it is possible to add an extra two 'electrons' to each ring which would not contribute to the charge so long as there is no spin-2 rotation to accept field momentum azimuthally or convey it. These additions would arise from two mutually orthogonal spin-2 linear vibrations, the analogue of spin-1 linear polarization. In fact, Leach *et al.* (2004) have generated beams with optical OAM having $l = 0$.[30] The four states in a ring could then be represented by two linear vibrations (charge-free), as well as a clockwise and a counterclockwise rotation (charged).

Considering a charge-free linear vibration in a nuclear ring, $m_o$ in (7.15) is replaced by $m'_o$ and it could serve as a neutron. So, a nuclear ring would possess two protons in opposite spin-2 rotations and two mutually perpendicular linearly vibrating neutrons, which together constitute an alpha particle. All four could participate in transitions between energy levels as the ring and its excited states exist independently of an external potential. In summary, a nuclear ring is an alpha particle when fully occupied.

## References

Bjerknes, V. F. K. *Fields of Force: A Course of Lectures in Mathematical Physics Delivered December 1 to 23*. New York: The Columbia University Press, 1906.

Bohr, N. 'On the constitution of atoms and molecules'. *The London, Edinburgh, and Dublin Philosophical Magazine*, 26 (1913a): 1–25.

---

[30] Optical OAM is equivalent to spin-2 in the present treatise.

Bohr, N. 'On the spectra of helium and hydrogen'. *Nature* October (1913b): 231–232.

Bohr, N. 'On the series spectrum of hydrogen and the structure of the atom'. *Philosophical Magazine*, 29 (1915): 332–335.

Bohr, N. 'Atomic structure'. *Nature*, 108 (1921): 208–209.

Bohr, N. 'The structure of the atom. Nobel lecture. December 11, 1922'. In H. Stephen (ed.), *The Dreams That Stuff Is Made Of*. Philadephia, PA: Running Press, 2011, pp. 104–147.

Clarke, B. R. *The Quantum Puzzle: Critique of Quantum Theory and Electrodynamics*. Singapore: World Scientific Publishing, 2017.

Dirac, P. *The Principles of Quantum Mechanics*, fourth edition. Oxford University Press, 1930; reprinted 2000.

Goudsmit, S., and G. E. Uhlenbeck, 'Opmerking over de spectra van waterstof en helium'. *Physica*, 5 (1925): 226–270.

Heisenberg, W., and P. Jordan. 'Anwendung dur Quantenmechanik auf das Problem der anomelen Zeemaneffeke'. *Zeitschrift für Physik*, 37 (1926): 263–277.

Horbatsch, M., and E. A. Hessels. 'Tabulation of the bound-state energies of atomic hydrogen'. *Physical Review A*, 93 (2016): 022513.

Johnson, R. C. *Atomic Spectra*. Reprinted second edition. London: Methuen, 1952.

Kragh, H. 'The fine structure of hydrogen and the gross structure of the physics community, 1916–26'. *Historical Study of the Natural Sciences*, 15 (1985): 67–127.

Langmuir, I. 'The arrangement of electrons in atoms and molecules'. *Journal of the American Chemical Society*, 41, 6 (1919): 868–934.

MacCullagh, J. 'An essay towards a dynamical theory of crystalline reflexion and refraction'. *Transactions of the Royal Irish Academy*. XXI (December 1839). In John H. Jellett and Samuel Haughton (eds.), *The Collected Works of James MacCullagh*. Dublin: Hodges, Figgis, & Co., 1880, pp. 145–184.

McMorran, B. J., A. Agrawal, I. M. Anderson, A. A. Herzing, H. J. Lezec, J. J. McClelland, and J. Unguris. 'Electron vortex beams with high quanta of orbital angular momentum'. *Science*, 331 (2011): 192–195.

Mehra, J., and H. Rechenberg. *The Historical Development of Quantum Theory*, Vol. 1, Part 1. New York: Springer–Verlag, 2001.

Michelson, A. A., and E. W. Morley. 'On a method of making the wave-length of sodium light the actual and practical standard of length'. *The London, Edinburgh, and Dublin Philosophical Magazine*, 24 (1887): 463–466.

Mohr, P. J., D. B. Newell, B. N. Taylor, and E. T. Tiesinga. 'Data and analysis for the CODATA 2017 special fundamental constants'. *Metrologia*, 55 (2018): 125–146.

Molina–Terriza, G., J. Recolons, J. P. Torres, L. Torner, and E. M. Wright. 'Observation of the dynamical inversion of the topological charge of an optical vortex'. *Physical Review Letters*, 87 (2001): 023902.

Nicholson, J. W. 'The constitution of the solar corona. II'. *Monthly Notices of the Royal Astronomical Society*, 72 (1912): 677–693.

Pauli, W. '*Über das Wasserstoffspektrum vom Standpunkt der neuen Quantenmechanik*'. *Zeitschrift für Physik*, 36 (1926): 336–363.

Sommerfeld, A. 'Zur quantentheorie der Spekrallinien'. *Annalen der Physik*, 51 (1916): 1–94.

Sommerfeld, A. *Atomic Structure and Spectral Lines*. London: Methuen, 1923.

Thomas, L. H. 'The motion of the spinning electron'. *Nature*, 117 (1926): 514.

Thomson, J. J. 'On the structure of the atom: an investigation of the stability and periods of oscillation of a number of corpuscles arranged at equal intervals around the circumference of a circle; with application of the results to the theory of atomic structure'. *Philosophical Magazine*, 7, 39 (1904): 237–265.

Uhlenbeck, G. E., and S. Goudsmit. 'Spinning electrons and the structure of spectra'. *Nature*, 117 (1926): 732–738.

Wilson, W. 'The quantum theory of radiation and the line spectra'. *The London, Edinburgh, and Dublin Philosophical Magazine*, 29 (1915): 795–802.

Wilson, W. 'The quantum of action'. *The London, Edinburgh, and Dublin Philosophical Magazine*, 31 (1916): 156–162.

## Chapter 9

# Hydrogen Atom Hyperfine Structure

*Using a Doppler shift, the fine-structure calculation has positioned the MVR frequency of a state at the mid-point of the two experimental hyperfine values by variation of the fine-structure constant multiplier Kalpha. The hyperfine structure calculation now varies the electron's spin-2 circuit eccentricity until the lower or upper hyperfine value for a state is obtained. The two spin-2 rotation senses allow both values to be calculated, one raising the mid-point value and the other lowering it. The minimum eccentricity, which corresponds to the greatest absorbed spin-2 field action, occurs at the major axis rotation angle of 135° with respect to the positive x axis for all levels. A simple hyperfine rule is also given that predicts the shift from the hyperfine mid-point value to the lower or upper value to at least the nearest MHz.*

## 9.1 Preliminary

Since the present MVR theory is not founded on quantum mechanics, there seems little point in setting out the history of the development of the hyperfine theory. So, the best course will be to continue along the lines of thought already set out and show how variation of the spin-2 eccentricity is able to produce exact MHz transition frequencies for the hyperfine levels of the hydrogen atom to 4 d.p.

The previous chapter has described how each elliptical ring, characterized by the Sommerfeld quantum numbers $(n_\phi, n_r)$, accommodates two levels. This was already known by Sommerfeld. For example, $(1, 1)$ has $2S_{1/2}$ and $2P_{1/2}$; see Table 8.2. Quantum mechanics uses the spin quantum number to distinguish between them: for $2S_{1/2}$, it is $Sp = 1/2$, and for $2P_{1/2}$, it is $Sp = -1/2$. However, since the basis of the present treatment does not rely on quantum mechanical considerations, this quantum number will be superfluous here. What distinguishes the levels in MVR theory is two different values of the fine-structure constant.

We have also assumed that prior to emission, every level collapses to the electron ground state for which the ring is circular. The electron ring is located at the oscillation boundary where it is at rest, and the advancing proton ring results in a Doppler shift of the emitted energy; see Section 8.3.8. With this in view, the fine-structure constant is modified by an automatic computer program search of its multiplier *Kalpha* until the fine-structure energy is aligned with the mid-point of the two experimental hyperfine energies for that level. It is at this point that the hyperfine calculation is introduced.

The hyperfine calculation addresses the way in which deviations from the mid-point of the hyperfine energies occur to produce the exact lower or upper energy. At the oscillation boundary, the electron OAM ring is stationary, but the proton ring is at its maximum speed and is advancing toward it. Consequently, there is a change in the proton magnetic momentum field (from its spin-2 circuit) cutting the electron spin-2 circuit. The spin-2 circuit for an electron in a field-free region is circular. However, when a change in field momentum is introduced, it results in a partial redistribution of the circular action in the electron spin-2 circuit into radial action so that the circuit becomes elliptical with an eccentricity $e$. This naturally affects the ability of the electron spin-2 circuit to take on spin-2 field momentum. So to maintain the invariance of the electron spin-2 action, the perimeter of the circuit is kept constant and equal to that of the circular case, which has radius $r_2$. This is essential since the electron mass depends on the perimeter in the spin-2 plane.

The proton spin-2 field will be assumed to be parallel and is set perpendicular to the line joining the proton–electron spin-2 centers; see Figure 9.1. Since the emissions occur from a point on the circular electron ground-state ring, the coordinates of a point on this ring fix this angle at 119.942° to the positive $x$ axis (common proton–electron axis). Although some justification for the choice of a parallel rather than a radial field is given, in an endeavor to keep to the main lines of the argument, this topic is not explored further here.

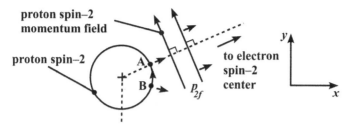

Figure 9.1   Proton spin-2 momentum field set perpendicular to the line joining the proton–electron spin-2 centers.

When the proton–electron spin-2 circuits rotate in the same sense, the advancing proton field causes a decrease in the electron spin-2 action, with an increase for the opposite sense. This results in either a decrease (same) or increase (opposite) in electron ring spin-3 momentum, which appears as a fine-structure energy adjustment. This is the origin of the negative or positive deviation from the aligned hyperfine energy mid-point in order to obtain the exact hyperfine values.

The program input angle $\phi$ is the orientation that the major axis of the electron spin-2 ellipse makes with the common proton–electron OAM ring axis, where the leftmost focus of the ellipse is at the electron spin-2 origin when $\phi = 0$; see Figure 9.3. Given the orientation $\phi$, the computer program searches for the value of the eccentricity that produces the exact hyperfine energy. By experimenting with the input value $\phi$, for most levels given in Table 9.1, the angle that produces the minimum eccentricity $e$ is $\phi = 135°$ (nearest degree).

Now, it could be argued that even the Sommerfeld theory of 1916 could be used to obtain the exact hyperfine energies simply by abandoning the accepted fine-structure constant and selecting a set of appropriate values that give the right results. With no rule for the selection of these alternative values, the calculation would then appear contrived. There would be some justification in this criticism since the MVR theory as presented is certainly in want of an explanation as to how *Kalpha* for a given state depends on the spin-3 radius of its electron ring. Nevertheless, the MVR theory differs from the Sommerfeld theory, indeed even quantum mechanics, in that it offers a definition of mass, an explanation of electric potential, a mechanism for charge difference, and an explanation of the magnetic momentum field.[1] There is also the idea of the self-potential, see (6.18), in which the spin-3 curvature appears to contribute to the mass. This surely has consequences for an atomic theory of gravitation.

Also, some interesting regularities occur. The minimum eccentricity required to produce the exact hyperfine energy occurs at $\phi = 135°$ for most of the levels given in Table 9.1.[2] In addition, using the aligned hyperfine mid-point, there is a simple hyperfine energy rule (Section 9.3.2) based on the effect the advancing proton field has on a circular electron spin-2 circuit ($e = 0$), giving a good agreement with the correct hyperfine energies. In fact, the calculation for the $1S_{1/2}$ state hyperfine energies gives exact agreement to four decimal places in the frequency (MHz). So, it is judged worthwhile to present these innovations to the reader in their imperfect state in the hope that a future insight will emerge that justifies the choice of these new fine-structure constants, characterized by the values of *Kalpha*.

## 9.2   Elements of the Calculation

### 9.2.1   *Parallel field assumption*

The line integral of a proton spin-2 momentum radial field (with curl zero) around an electron spin-2 circuit would produce a null result. There could be no hyperfine interaction. However, the field that will be posited here is to be perpendicular to the line joining the proton–electron spin-2 centers, as a parallel field; see Figure 9.1. This was the assumption made in a previous work to derive the Lorentz force from the unloaded-ring MVR theory (Clarke 2017, 271, 284–296). Both the

---

[1] That is, the vector potential which becomes a magnetic momentum on multiplication by the charge.

[2] For $4D_{5/2}$, the minimum $e$ occurs at $\phi = 127.9°$. There is no $e$ that works at $135°$ for this level.

proton and electron spin-2 profiles are elliptical.[3] The former is the source of the proton magnetic momentum field.

The postulate of a parallel field has some justification. During the SAM cluster's progress from position $B$ along the proton's spin-2 circuit (see Figure 9.1), the effect on the surrounding space of the SAM cluster at point $B$ occurs before the effect at point $A$. This means that the radial transmission of the field from $B$ occurs before point $A$ and this tends to straighten out the curvature of the arc. Apart from this brief comment, the generation of the parallel field will not be examined in any significant detail here. It will only be presented as a postulate based on reasonable grounds. Neither will the effect of the electron field on the proton spin-2 momentum circuit be analyzed, mainly because the effect on the proton field-free circular circuit should be so weak as to insignificantly alter its eccentricity from $e = 0$.

### 9.2.2   *Proton spin-2 field momentum vector*

We now move to the formulation of the proton spin-2 field momentum vector. Figure 9.2 shows the proton field momentum $p'_{2f}$ as a

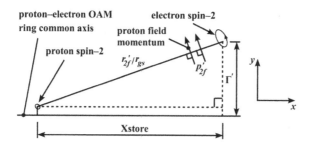

Figure 9.2   Proton–electron spin-2 arrangement in the $1S_{1/2}$ state at the end of the bound-state oscillation, with a parallel proton momentum field in the vicinity of the electron spin-2 ring. The distances $Xstore, \Gamma'$, and the proton spin-2 radius $r'_{2f}$ are each expressed as a fraction of the electron ground-state radius $r_{gs}$.

---

[3] As an approximation, the proton spin-2 circuit has been assumed to be circular here. The effect of the relative motion of the electron and proton, together with the much weaker strength of the electron spin-2 field momentum, should not significantly distort the proton field-free circular spin-2 circuit.

parallel structure perpendicular to the line joining the proton–electron spin-2 centers. All distances are now expressed as a fraction of the electron ground-state spin-3 radius $r_{gs}$, $Xstore = x/r_{gs}$ and $\Gamma' = (r_3 - r_3')/r_{gs} = (gamma - M)$.[4] Let us consider the counterclockwise field that emanates from the proton spin-2 circuit. To set up the appropriate vector, we need to see the entire proton–electron arrangement in order to get the correct components. The proton field momentum $p_{2f}'$ is perpendicular to the radius $r_{2f}'$ so that the vector is given by

$$\vec{p}_{2f}' = \frac{\sqrt{2}\hbar}{Mr_{2f}'} \begin{pmatrix} -\frac{(y-r_3')}{r_{2f}'} \\ \frac{x}{r_{2f}'} \\ 0 \end{pmatrix} \tag{9.1}$$

where, from Figure 6.7, we have

$$r_{2f}' = ((x + r\cos\gamma)^2 + (y - r_3' + r\sin\gamma)^2)^{1/2} \tag{9.2}$$

The presence of $\sqrt{2}/M$ in (9.1) is justified as follows. Both the electron and proton spin-3 rings rotate at the same speed $v_3 = v_3'$ in field-free regions; see (7.15). However, the proton OAM circular ring is the fraction $M = m/m'$ of the electron ring circumference. So, there are $M^{-1}$ presentations of proton field momentum in one electron SAM cluster spin-3 revolution. In addition, the electron at its oscillation boundary has only $1/\sqrt{2}$ of its field-free ring speed; see (7.16). This creates an additional factor of $\sqrt{2}$.

### 9.2.3    *Electron spin–2 circuit*

The next task is to parameterize the electron spin-2 circuit. For the hydrogen atom, the action around one circuit of a circular profile is $h$, which is given purely to the azimuthal direction. For an elliptical profile, this quantity of action is to be retained, but, due to the external field intervention, it is now distributed between the radial and azimuthal directions. To preserve this electron action,

---

[4]For the electron ground state $gamma = 1$.

we demand that the perimeter of the ellipse remains equal to the circular circumference $2\pi r_2$, where $r_2$ is the radius of the ground-state circular field-free spin-2 circuit. In fact, the definition of mass is tied up with the perimeter length; see (7.26). We can write the elliptical perimeter as

$$perimeter = 2\pi a \times perim \qquad (9.3)$$

where $a$ is the major axis. To ensure that the perimeter remains constant over all eccentricities, we impose the condition that

$$a = \frac{r_2}{perim} \qquad (9.4)$$

For a circle and eccentricity $e = 0$, we have $perim = 1$. Cayley (1876, 54) has given a formula to calculate $perim$ provided that $a/b$ is large ($e$ close to unity); see also Chandrupatla and Osler (2010, 128). This can be stated as

$$
\begin{aligned}
perim = 1 &+ \frac{1}{2}\left(\ln\left(\frac{4}{k}\right) - \frac{1}{1\times 2}\right)k^2 \\
&+ \frac{1^2 \times 3}{2^2 \times 4}\left(\ln\left(\frac{4}{k}\right) - \frac{2}{1\times 2} - \frac{1}{3\times 4}\right)k^4 \\
&+ \frac{1^2 \times 3^2 \times 5}{2^2 \times 4^2 \times 6}\left(\ln\left(\frac{4}{k}\right) - \frac{2}{1\times 2} - \frac{2}{3\times 4} - \frac{1}{5\times 6}\right)k^6 + \cdots
\end{aligned}
$$
$$(9.5)$$

where $k^2 = (1 - e^2)$. This calculation appears in lines 501–511 in the program to order $k^{18}$ and is intended for $e > 0.9$. In fact, there is series convergence to 5 d.p. for $e = 0.8$, to 6 d.p for $e = 0.85$, to 8 d.p for $e = 0.9$, to 9 d.p. for $e = 0.92$, and to 10 d.p. for $e = 0.95$. The alternative calculation in lines 513–533 is based on the MacLaurin expansion, which works better the closer $e$ is to zero (Chandrupatla and Osler 2010, 122), but the computer program takes longer to run than the Cayley case. To obtain MacLaurin convergence to 16 d.p., we need terms to order $e^{64}$ for $e = 0.7$; order $e^{108}$ for $e = 0.8$; order $e^{136}$ for $e = 0.85$; and order $e^{204}$ for $e = 0.9$. So, this calculation is only used in the program for $e \le 0.9$ to order $e^{204}$.

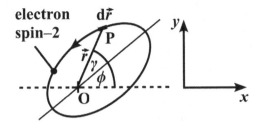

Figure 9.3    The electron spin-2 ellipse with major axis at angle $\phi$ to the $x$ axis (see Figure 9.2), and radius OP at angle $\gamma$ to the major axis.

So, to summarize, for $e \leq 0.9$, the MacLaurin calculation is used. It is slower, but we can get accuracy to 16 d.p. For $e > 0.9$, we switch to the Cayley calculation. It is much quicker, but we sacrifice accuracy, which can be obtained to at least 8 d.p.

So, now we give the coordinates for an element on the electron spin-2 ellipse perimeter relative to the proton spin-2 axis. Figure 9.3 shows the enlarged electron spin-2 ellipse from Figure 9.2, albeit with a different orientation.

In respect of Figure 9.3, and using (9.4), we have

$$\hat{r} = \begin{pmatrix} \cos(\gamma + \phi) \\ \sin(\gamma + \phi) \\ 0 \end{pmatrix}$$

$$r = \frac{a(1 - e^2)}{1 - e \cos \gamma} = \frac{r_2(1 - e^2)}{perim(1 - e \cos \gamma)}$$

(9.6)

The angle $\phi$ is that through which the major axis of the ellipse is rotated counterclockwise with respect to the positive $x$ axis, and is adjustable as the third input in the program. Consequently, by differentiating with respect to $\gamma$, and only considering the azimuthal component for spin-2, we have

$$d\hat{r}_\gamma = \begin{pmatrix} -\sin(\gamma + \phi) \\ \cos(\gamma + \phi) \\ 0 \end{pmatrix} d\gamma$$

(9.7)

### 9.2.4 *Electron and proton speeds*

There are two speeds that will enter into the calculation: the speed $v'_x$ of the proton along the $x$ axis with respect to the stationary (in the center of mass frame) electron OAM ring, and the speed $v_3$ of the electron SAM cluster in the OAM ring at the oscillation boundary (it has no speed along the $x$ axis). The former is the speed at which the proton field is approaching the stationary electron OAM ring when the latter is at the oscillation boundary, and it participates in the calculation for the change in the proton momentum field. The latter is taken from (7.17) so that

$$v_3 = \frac{\alpha c}{\sqrt{2}Y\left(1 + \frac{\alpha^2}{Y^2}\right)^{1/2}} \tag{9.8}$$

and the former is

$$v'_x = \frac{dx}{dt} = -\frac{M\alpha c}{\sqrt{2}Y\left(1 + \frac{\alpha^2}{Y^2}\right)^{1/2}} \tag{9.9}$$

This is negative due to the relative proton–electron displacement $x$ decreasing as the proton ring approaches the stationary electron ring. The presence of $M$ arises from conservation of linear momentum considerations set out in Section 6.5.2.

### 9.2.5 *Change in electron spin-2 momentum*

As the proton ring advances an incremental distance $\Delta x$ toward the stationary electron ring along the $x$ axis, the intruding field on an increment $\Delta \hat{r}_\gamma$ on the electron spin-2 circuit varies. The aim of the calculation is to calculate the total change in electron spin-2 momentum $\Delta p$ from a line integral around the electron azimuthal circuit during the electron spin-2 time period $T_2$. Since its spin-2 action is to be conserved, there is a displacement into spin-3 momentum in the electron OAM ring. Depending on the relative rotation senses of the proton and electron spin-2, this either adds to or subtracts from the electron spin-3 fine-structure energy.[5] To convert the change in spin-2 momentum $\Delta p$ to a change in spin-3 energy as a hyperfine correction $\Delta\varepsilon_{hf}$, it needs to be multiplied by the electron speed $v_3$ in the OAM ring. Thus,

$$\Delta\varepsilon_{hf} = \Delta p \cdot v_3 = \frac{v_3}{2\pi} \int_0^{2\pi} \frac{d\vec{p}'_{2f}}{dx} \cdot d\hat{r}_\gamma \int_0^{T_2} \frac{dx}{dt} dt \qquad (9.10)$$

The division by $2\pi$ results from a mean value calculation of $\Delta p$ taken over the angle $\gamma$. So, using (9.1), (9.7), and (9.9), the calculation becomes

$$\Delta\varepsilon_{hf} = -\frac{v_3 v'_x T_2}{2\pi} \frac{\sqrt{2}\hbar}{M} \int_0^{2\pi} \frac{d}{dx} \left( r'^{-2}_{2f} \left( \begin{array}{c} -(y - r'_3) \\ x \\ 0 \end{array} \right) \right)$$

$$\cdot \left( \begin{array}{c} -\sin(\gamma + \phi) \\ \cos(\gamma + \phi) \\ 0 \end{array} \right) d\gamma \qquad (9.11)$$

which since $\hbar = m_o c r_2$ — see (6.3) — and $T_2 = 2\pi r_2 \left(1 + \frac{\alpha^2}{Y^2}\right)^{1/2}/c$, then since $v'_x = dx/dt$ does not significantly vary during the electron spin-2 time period $T_2$, using (9.8) and (9.9), we have

$$\Delta\varepsilon_{hf} = -\frac{\sqrt{2} m_o \alpha^2 c^2 r_2^2}{2Y^2 \left(1 + \frac{\alpha^2}{Y^2}\right)^{1/2}} \int_0^{2\pi} \frac{d}{dx} \left( r'^{-2}_{2f} \left( \begin{array}{c} -(y - r'_3) \\ x \\ 0 \end{array} \right) \right)$$

$$\cdot \left( \begin{array}{c} -\sin(\gamma + \phi) \\ \cos(\gamma + \phi) \\ 0 \end{array} \right) d\gamma \qquad (9.12)$$

---

[5]For a decreasing proton–electron displacement, same-sense rotations produce a decrease in electron spin-2 momentum, while opposite senses give an increase.

Taking account of (9.2), the differentiated part of the integrand is

$$\frac{d}{dx}\left(r_{2f}'^{-2}\begin{pmatrix}-(y-r_3')\\x\\0\end{pmatrix}\right)\cdot\begin{pmatrix}-\sin(\gamma+\phi)\\\cos(\gamma+\phi)\\0\end{pmatrix}$$

$$=\frac{1}{X^2}\left(1+\frac{r^2}{X^2}+\frac{2r\left(x\cos(\gamma+\phi)+(y-r_3')\sin(\gamma+\phi)\right)}{X^2}\right)^{-1}$$

$$\times\cos(\gamma+\phi)-\frac{2\left(x+r\cos(\gamma+\phi)\right)}{X^4}$$

$$\times\left(1+\frac{r^2}{X^2}+\frac{2r\left(x\cos(\gamma+\phi)+(y-r_3')\sin(\gamma+\phi)\right)}{X^2}\right)^{-2}$$

$$\times\left((y-r_3')\sin(\gamma+\phi)+x\cos(\gamma+\phi)\right) \tag{9.13}$$

where $X^2 = x^2 + (y-r_3')^2$. We now give an approximation of the result. For the circular field-free spin-2 case, when $e = 0$ in (9.6), then $perim = 1$ and $r = r_2$. A binomial expansion to first order in $r_2$ gives (9.13) as

$$\frac{d}{dx}\left(r_{2f}'^{-2}\begin{pmatrix}-(y-r_3')\\x\\0\end{pmatrix}\right)\cdot\begin{pmatrix}-\sin(\gamma+\phi)\\\cos(\gamma+\phi)\\0\end{pmatrix}$$

$$=\frac{1}{X^2}\left(1+\frac{r_2^2}{X^2}-\frac{2r_2\left(x\cos(\gamma+\phi)+(y-r_3')\sin(\gamma+\phi)\right)}{X^2}\right)$$

$$\times\cos(\gamma+\phi)-\frac{2\left(x+r_2\cos(\gamma+\phi)\right)}{X^4}$$

$$\times\left(1+\frac{r_2^2}{X^2}-\frac{4r_2\left(x\cos(\gamma+\phi)+(y-r_3')\sin(\gamma+\phi)\right)}{X^2}\right)$$

$$\times\left((y-r_3')\sin(\gamma+\phi)+x\cos(\gamma+\phi)\right) \tag{9.14}$$

As an approximation, with $e = 0$, the orientation become irrelevant, so we set $\phi = 0$. Only products of even powers of $\sin(\gamma)$

and $\cos(\gamma)$ contribute to the integral (9.12). So,

$$\int_0^{2\pi} \frac{d}{dx}\left(r_{2f}^{\prime-2}\begin{pmatrix} -(y-r_3') \\ x \\ 0 \end{pmatrix}\right) \cdot \begin{pmatrix} -\sin(\gamma) \\ \cos(\gamma) \\ 0 \end{pmatrix} d\gamma \sim \frac{4\pi r_2 x}{X^4} \quad (9.15)$$

Together with (9.12), we arrive at

$$\Delta\varepsilon_{hf} \sim -\frac{m_o\alpha^2 c^2}{2\left(1 - \frac{\alpha^2}{2Y^2\left(1+\frac{\alpha^2}{Y^2}\right)}\right)^{1/2} Y^2\left(1+\frac{\alpha^2}{Y^2}\right)}\left(\frac{4\sqrt{2}\pi r_2^3 x}{X^4}\right)$$

$$\times \left(1 + \frac{\alpha^2}{Y^2}\right)^{1/2}\left(1 - \frac{\alpha^2}{2Y^2\left(1+\frac{\alpha^2}{Y^2}\right)}\right)^{1/2} \quad (9.16)$$

Apart from the reduced mass, which has yet to be introduced, (9.12) has been modified to (9.16) to include the fine-structure energy form; see (9.18).The first of the four expressions in the product is the fine-structure energy. So, the rest of the terms constitute the hyperfine adjustment. This is the approximate result for same-sense rotation of the proton and electron spin-2 circuits with the distance between the proton and electron decreasing; that is, the proton momentum field is increasing in strength.

### 9.2.6    *Programming the hyperfine adjustment*

The full numerical integral calculation using (9.12) and (9.13) appears in the computer program in lines 535–558. For the program, all distances are to be divided by the electron ground-state radius $r_{3,gs} = r_{gs}$, so that $Xstore = x/r_{gs}$, $\Gamma = y/r_{gs}$, $M = m/m' = r_2'/r_{gs}$, see (A.15), and $\Gamma' = (gamma - M)$. The variable names $Xstore$ and $gamma$ are used in the computer program. For the electron ground state $gamma = 1$. Also, $r_2/r_{gs} = \alpha/(1-\alpha^2)^{1/2}$, having used the loaded-ring version of (A.3). This leaves the approximation (9.12) and (9.13) as

$$\varepsilon_{tot} = \varepsilon_{fs}(1 \mp \delta_{hf}) \tag{9.17}$$

where the negative sign is for the same-sense proton–electron spin-2 rotation and a decreasing displacement between the two rings. Here, after introducing the reduced mass, we have

$$\varepsilon_{fs} = \frac{m_o \alpha^2 c^2}{2(1 + M)\left(1 - \frac{\alpha^2}{2Y^2\left(1 + \frac{\alpha^2}{Y^2}\right)}\right)^{1/2} Y^2 \left(1 + \frac{\alpha^2}{Y^2}\right)} \tag{9.18}$$

and

$$\delta_{hf} = \frac{\sqrt{2}\alpha^2}{1 - \alpha^2}\left(1 + \frac{\alpha^2}{Y^2}\right)^{1/2}\left(1 - \frac{\alpha^2}{2Y^2\left(1 + \frac{\alpha^2}{Y^2}\right)}\right)^{1/2}$$

$$\times \int_0^{2\pi} \left[ \frac{1}{\left(Xstore^2 + \Gamma'^2\right)}\left(1 + \frac{r^2}{\left(Xstore^2 + \Gamma'^2\right)}\right.\right.$$

$$\left. + \frac{2r\left(Xstore\cos\left(\gamma + \phi\right) + \Gamma'\sin\left(\gamma + \phi\right)\right)}{\left(Xstore^2 + \Gamma'^2\right)}\right)^{-1}\cos\left(\gamma + \phi\right)$$

$$- \frac{2\left(Xstore + r\cos\left(\gamma + \phi\right)\right)}{\left(Xstore^2 + \Gamma'^2\right)^2}\left(1 + \frac{r^2}{\left(Xstore^2 + \Gamma'^2\right)}\right.$$

$$\left. + \frac{2r\left(Xstore\cos\left(\gamma + \phi\right) + \Gamma'\sin\left(\gamma + \phi\right)\right)}{\left(Xstore^2 + \Gamma'^2\right)}\right)^{-2}$$

$$\times \left(\Gamma'\sin\left(\gamma + \phi\right) + Xstore\cos\left(\gamma + \phi\right)\right) \Bigg] d\gamma \tag{9.19}$$

where $\Gamma' = 1 - M$ for the ground state. From (9.6), the radius $r$ is given by

$$r = \frac{\alpha(1 - e^2)}{perim\,(1 - e\cos\gamma)\,(1 - \alpha^2)^{1/2}} \tag{9.20}$$

where $r_2/r_{gs} = \alpha/(1 - \alpha^2)^{1/2}$. In (9.16), once the reduced mass has been included, the last three bracketed terms in the product give the approximate value $7.479\,189\,067 \times 10^{-7}$, against the computer program's accurate numerical integration over all orders of $\alpha$, which gives $7.479\,388\,245 \times 10^{-7}$.[6]

## 9.3  Hyperfine Results

### 9.3.1  *Spin–2 major axis and eccentricity*

In Table 9.1, we show the results from the computer program that produces exact hyperfine energies for the hydrogen atom to

Table 9.1  Selection of levels with electron spin–2 set at '+' (same sense as proton).

| Level | *Kalpha* | *e* | *a* |
|---|---|---|---|
| $1S_{1/2}$ | 0.999 997 601 435 212 8 | 0.911 642 714 018 701 4 | 1.357 629 600 039 600 0 |
| $2S_{1/2}$ | 0.999 998 784 659 971 4 | 0.959 189 058 068 242 7 | 1.443 879 799 547 958 5 |
| $2P_{1/2}$ | 0.999 999 410 077 344 9 | 0.984 976 374 007 663 1 | 1.510 741 198 268 423 9 |
| $2P_{3/2}$ | 0.999 999 387 451 329 7 | 0.992 187 731 182 009 5 | 1.535 175 239 368 075 4 |
| $3S_{1/2}$ | 0.999 999 186 941 079 1 | 0.972 557 514 774 115 4 | 1.475 682 097 723 238 0 |
| $3P_{1/2}$ | 0.999 999 605 863 149 8 | 0.988 976 218 679 183 1 | 1.523 775 092 355 692 4 |
| $3P_{3/2}$ | 0.999 999 590 780 765 7 | 0.993 822 338 695 106 7 | 1.541 403 460 455 534 3 |
| $3D_{3/2}$ | 0.999 999 597 598 818 8 | 0.995 169 136 588 970 9 | 1.546 816 723 815 535 3 |
| $3D_{5/2}$ | 0.999 999 593 758 395 4 | 0.995 919 361 847 501 8 | 1.549 968 109 789 841 9 |
| $4S_{1/2}$ | 0.999 999 389 317 118 6 | 0.978 865 709 614 896 8 | 1.492 644 450 350 815 8 |
| $4P_{1/2}$ | 0.999 999 704 090 174 5 | 0.990 975 923 678 784 8 | 1.530 757 086 458 565 6 |
| $4P_{3/2}$ | 0.999 999 692 778 991 5 | 0.994 658 496 310 976 5 | 1.544 730 249 513 492 9 |
| $4D_{3/2}$ | 0.999 999 697 985 676 6 | 0.995 691 133 294 563 6 | 1.548 997 801 884 372 1 |
| $4D_{5/2}$ | 0.999 999 695 105 233 1 | 0.997 582 607 793 420 8 | 1.557 416 109 411 496 7 |
| $5S_{1/2}$ | 0.999 999 511 083 889 1 | 0.982 550 251 129 592 2 | 1.503 326 716 954 031 5 |
| $5P_{1/2}$ | 0.999 999 763 130 696 6 | 0.992 184 359 644 229 0 | 1.535 162 727 698 462 8 |

*Notes*: The *Kalpha* (fine-structure constant multiplier), spin–2 eccentricity $e$, and major axis $a$ (as a multiplier of the circular radius $r_2$) are given. All optimum rotation angles $\phi$ (where $e$ is minimum) are at 135° (to the nearest degree) except level $4D_{5/2}$, which requires an input angle of 127 degrees. A Doppler blue shift is selected. The program outputs are (24), (25), and (16).

---

[6] The computer program gives the result for $\delta_{hf}$ as *hypercalc* (line 558), which converges at *thlimit* = 100 (line 538) to 15 decimal places.

4 d.p. (MHz). The proton and electron spin-2 senses have been selected to be the same (+). The key values are as follows:

(1) The fine-structure constant multiplier *Kalpha* that brings the fine-structure energy into alignment with the mid-point of the experimental hyperfine values, computer output (24).
(2) The minimum eccentricity $e$ of the electron spin-2 elliptical circuit, computer output (25). This occurs at $\phi = 135°$ (to the nearest degree), apart from $4D_{5/2}$ where the input rotation angle $\phi = 127°$.
(3) The length of the major axis $a$, as a multiplier of the spin-2 circular radius $r_2$, under the condition that the elliptical perimeter retains the same length $2\pi r_2$ as the circular case, computer output (16).

### 9.3.2 *The hyperfine rule*

At this point, we present an interesting rule that gives good agreement with the deviation of the upper or lower hyperfine frequency (MHz) from the mid-point of the two values. As we have already seen, for the ground state of hydrogen with the spin-2 eccentricity set at $e = 0$, the deviation $\delta_{hf}$ taken from (9.17)–(9.19) has the value $7.479\,388\,245 \times 10^{-7}$ obtained from a numerical integration. The rule we now posit is as follows:

$$\delta_{hf} = \frac{7.479\,388\,245 \times 10^{-7} \times hypconst}{2\sqrt{3}Y} \qquad (9.21)$$

or

$$\delta_{hf} = \frac{2.159\,113\,408 \times 10^{-7} \times hypconst}{Y} \qquad (9.22)$$

where $Y = n_r + (n_\phi^2 - \alpha^2)^{1/2}$. The values of *hypconst* are set out in Table 9.2. Table 9.3 shows how well this rule gives correspondence with the experimental hyperfine energies.

Of course, $\alpha$ in $Y$ is to be adjusted using *Kalpha*. In the computer program, this calculation appears only in the print routine [*printout2*] in lines 907, 916, and 925. We can now understand from

Table 9.2   The hyperfine rule
(9.23) with *hypconst* values.

| States | *hypconst* in **(9.21)** |
|--------|--------------------------|
| $nS_{1/2}$ | 1 |
| $nP_{1/2}$ | 1/3 |
| $nP_{3/2}$ | 2/15 |
| $nD_{3/2}$ | 2/25 |
| $nD_{5/2}$ | 50/972 |

Table 9.3   Selection of lower and upper hyperfine frequencies (MHz) predicted by the hyperfine rule (9.22) compared with experimental values (Horbatsch and Hessels 2016).

| | Lower hyperfine freq. (MHz) | Upper hyperfine frequency (MHz) |
|--------|------------------------------|----------------------------------|
| $1S_{1/2}$ rule | 3 288 086 502.258 9 | 3 288 087 921.167 3 |
| expt | 3 288 086 502.010 2 | 3 288 087 922.416 0 |
| $2S_{1/2}$ rule | 822 025 399.570 3 | 822 025 577.057 3 |
| expt | 822 025 399.535 4 | 822 025 577.092 2 |
| $2P_{1/2}$ rule | 822 026 486.969 9 | 822 026 546.132 4 |
| expt | 822 026 486.966 4 | 822 026 546.135 9 |
| $2P_{3/2}$ rule | 822 015 523.838 7 | 822 015 547.503 1 |
| expt | 822 015 523.845 1 | 822 015 547.496 7 |
| $3S_{1/2}$ rule | 365 343 565.305 4 | 365 343 617.893 8 |
| expt | 365 343 565.294 9 | 365 343 617.904 3 |
| $3P_{1/2}$ rule | 365 343 888.940 4 | 365 343 906.469 8 |
| expt | 365 343 888.939 2 | 365 343 906.471 0 |
| $3P_{3/2}$ rule | 365 340 640.602 1 | 365 340 647.613 8 |
| expt | 365 340 640.604 0 | 365 340 647.611 9 |
| $3D_{3/2}$ rule | 365 340 646.986 3 | 365 340 651.193 3 |
| expt | 365 340 646.986 5 | 365 340 651.193 1 |
| $3D_{5/2}$ rule | 365 339 564.099 4 | 365 339 566.804 6 |
| expt | 365 339 564.100 3 | 365 339 566.803 7 |
| $4S_{1/2}$ rule | 205 505 287.762 4 | 205 505 309.948 0 |
| expt | 205 505 287.757 9 | 205 505 309.952 5 |
| $4P_{1/2}$ rule | 205 505 424.534 2 | 205 505 431.929 4 |
| expt | 205 505 424.533 7 | 205 505 431.929 9 |

*Note*: The lower value requires the input '+' while the upper value needs '−'.

Table 9.4   *Kalpha* values without
a Doppler effect for the first four
circular states.

| State | Kalpha |
|-------|--------|
| $1S_{1/2}$ | 0.999 998 818 524 |
| $2P_{3/2}$ | 0.999 999 995 998 |
| $3D_{5/2}$ | 0.999 999 999 456 |
| $4F_{7/2}$ | 0.999 999 999 922 |

this analysis why an eccentricity $e > 0$ is required. By inspecting (9.21), we can see that the proton magnetic field momentum that inhabits the electron *circular* spin-2 circuit ($e = 0$) needs to be reduced by a factor $\sim 2\sqrt{3}$ to obtain the value in (9.22). Increasing the eccentricity of the electron spin-2 circuit achieves this by allowing less of its azimuthal action to be exposed to the intruding proton momentum field.

An alternative model would be to abandon the Doppler shift so that states emit from their own rings (rather than collapse to the ground state). Then the following *Kalpha* values (Table 9.4) bring the fine structure energies into alignment with the average of the two hyperfine values. Emissions would then occur perpendicular to the proton–electron axis along which the proton motion occurs.

## References

Cayley, A. *An Elementary Treatise on Elliptic Functions.* Cambridge: Deighton, Bell, and Co., 1876.

Chandrupatla, T. R., and T. J. Osler. 'The perimeter of an ellipse'. *Mathematical Scientist*, 35 (2010): 122-131.

Clarke, B. R. *The Quantum Puzzle: Critique of Quantum Theory and Electrodynamics.* World Scientific Publishing, 2017.

Horbatsch, M., and E. A. Hessels. 'Tabulation of the bound-state energies of atomic hydrogen'. *Physical Review A*, 93 (2016): 022513.

Kramida, A. E. 'A critical comparison of experimental data on spectral lines and energy levels of hydrogen, deuterium, and tritium'. *Atomic Data and Nuclear Tables*, 96 (November 2010): 586–644.

# Appendix A

# The Unloaded OAM Ring

## A.1  The Average Spin-3 Radius $\bar{R}_3$

We have seen that the HSD takes on helical trajectories; see Figure 6.2(a). The spin-2 and spin-3 speeds on the OAM ring, and in the field in general, are to be invariant. The spin-2 radius $R_2$ for the unloaded ring is also constant, so with constant speed the corresponding angular speed $\dot{\gamma}$ must also be fixed. However, as the spin-3 radius $R_{3f}$ varies over the spin-2 circuit, $R_3 - R_2 \leq R_{3f} \leq R_3 + R_2$, if the speed is to remain at $\alpha c$, the corresponding angular speed $\omega$ must vary to compensate. Thus,

$$\begin{aligned} R_2\dot{\gamma} &= c(1 - \alpha^2)^{1/2} \\ R_{3f}\omega &= \alpha c \end{aligned} \tag{A.1}$$

where $\dot{\gamma}$ is the spin-2 angular speed. Also, we must have

$$\begin{aligned} v_2 R_2 &= c(1 - \alpha^2)^{1/2} R_2 = \frac{\hbar}{m_{2u}} \\ v_3 \bar{R}_3 &= \alpha c \bar{R}_3 = \frac{\hbar}{m_{3u}} \end{aligned} \tag{A.2}$$

where (6.6) gives $m_{2u} = m_{2o}(1 - \alpha^2)^{-1}$.[1] If spin-3 mass $m_{3u}$ is to have any meaning, we set it equal to the spin-2 mass $m_{2u}$ given by (6.5) and (6.6), and proceed to calculate the average spin-3 radius $\bar{R}_3$ at which this occurs. It follows from (A.2) that

$$\frac{R_2}{\bar{R}_3} = \frac{\alpha}{(1 - \alpha^2)^{1/2}} \tag{A.3}$$

---

[1] For the loaded mass ring, we have (7.6).

Referring to Figure 6.4, the variable radius $R_{3f}$ is taken with respect to the OAM ring origin and terminates at a point on the spin-2 circuit. This is computed from

$$R_{3f} = ((R_3 + R_2 \sin \gamma)^2 + (R_2 \cos \gamma)^2)^{1/2}$$
$$= (R_2^2 + R_3^2 + 2R_2 R_3 \sin \gamma)^{1/2} \tag{A.4}$$

### A.1.1    Calculation of $\bar{R}_3$

Let $Q = 1 + R_2^2/R_3^2$. Then, (A.4) can be recast as

$$R_{3f} = Q^{1/2} R_3 \left( 1 + \frac{2R_2 \sin \gamma}{QR_3} \right)^{1/2} \tag{A.5}$$

We now set up a calculation for the average spin-3 radius $\bar{R}_{3f}$ taken around the spin-2 circuit thus

$$\bar{R}_3 = \frac{Q^{1/2} R_3}{2\pi} \int_0^{2\pi} \left( 1 + \frac{2R_2 \sin \gamma}{QR_3} \right)^{1/2} d\gamma \tag{A.6}$$

Considering only the even powers of the $\sin \gamma$ terms that make a contribution to the integration, we have

$$\bar{R}_3 = \frac{Q^{1/2} R_3}{2\pi} \int_0^{2\pi} \left( 1 - \frac{R_2^2 \sin^2 \gamma}{2R_3^2 Q^2} - \frac{5R_2^4 \sin^4 \gamma}{8R_3^4 Q^4} \right.$$
$$\left. - \frac{21 R_2^6 \sin^6 \gamma}{16 R_3^6 Q^6} + \cdots \right) d\gamma \tag{A.7}$$

After integrating, expanding the powers of $Q$, and collecting like terms, we arrive at

$$\bar{R}_3 = R_3 \left( 1 + \frac{R_2^2}{4R_3^2} + \frac{R_2^4}{64R_3^4} + \frac{R_2^6}{256R_3^6} + \cdots \right) \tag{A.8}$$

## A.1.2  *Calculation of* $\mathbf{R_2/R_3}$

From (A.3) and (A.8), we find

$$\frac{R_2}{R_3} = \frac{\bar{R}_3}{R_3}\frac{R_2}{\bar{R}_3} = \frac{\alpha}{(1-\alpha^2)^{1/2}}\left(1 + \frac{R_2^2}{4R_3^2} + \frac{R_2^4}{64R_3^4} + \frac{R_2^6}{256R_3^6} + \cdots\right)$$

(A.9)

Ignoring terms higher than second order in $R_2/R_3$, we can find the following approximate solution:

$$\frac{R_2}{R_3} \sim \frac{\alpha}{(1-\alpha^2)^{1/2}} + \frac{\alpha^3}{(1-\alpha^2)^{3/2}}$$

(A.10)

The variable $R_3$ is the radius of the axis of the spin-2 rotation.

## A.1.3  *Mass and ring radii*

From (6.7), we have for the electron

$$\hbar = m_{2u}c(1-\alpha^2)^{1/2}R_2$$

(A.11)

and similarly for the proton (primed variables) which is merely a scaled-down version of the electron ring

$$\hbar = m'_{2u}c(1-\alpha^2)^{1/2}R'_2$$

(A.12)

A division yields

$$\frac{m_{2u}}{m'_{2u}} = \frac{R'_2}{R_2}$$

(A.13)

Equation (6.6) then allows us to form a ratio of proton to electron unloaded rest mass

$$\frac{m_{2o}}{m'_{2o}} = \frac{R'_2}{R_2}$$

(A.14)

From (A.3) and (A.10) and their primed proton counterparts, we can extend (A.14) as

$$M = \frac{m_{2o}}{m'_{2o}} = \frac{m}{m'} = \frac{R'_2}{R_2} = \frac{R'_3}{R_3} = \frac{\bar{R}'_3}{\bar{R}_3} \qquad (A.15)$$

So, all OAM mass rings are identical in construction and differ only by a scale factor that is identical to that of their rest masses.

# Appendix B

# Sommerfeld Fine Structure

## B.1 Sommerfeld Mass, Radius, and Action

This section refers to the discussion in Section 7.2.1. The starting point for Sommerfeld is to add an external potential energy term to the total mass on the left in (B.1) and add squares of azimuthal and radial momenta to the right side:

$$\left(mc + \frac{Z\hbar\alpha}{r_{3f}}\right)^2 = (m_oc)^2 + \left(\frac{n_\phi\hbar}{r_{3f}}\right)^2 + (p_r)^2 \tag{B.1}$$

Rearrangement gives

$$p_r = \left(\frac{\hbar^2(Z^2\alpha^2 - n_\phi^2)}{r_3^2} + \frac{2m_o\hbar Z\alpha c}{r_3} + (m^2 - m_o^2)c^2\right)^{1/2} \tag{B.2}$$

We now apply the radial action condition

$$\oint p_r dr_3 = n_r h \tag{B.3}$$

and the result using (7.5) is

$$m = \frac{m_o}{\left(1 + \frac{Z^2\alpha^2}{Y^2}\right)^{1/2}} \tag{B.4}$$

where $Y = n_r + (n_\phi^2 - Z^2\alpha^2)^{1/2}$. If we now select the azimuthal case $n_r = 0$, then (B.2) is solved for $r_3$ by setting the discriminant to zero. The solution to this is

$$m^2 - m_o^2 = -\frac{m_o^2 Z^2 \alpha^2}{n_\phi^2\left(1 - \frac{Z^2\alpha^2}{n_\phi^2}\right)} \tag{B.5}$$

Substituting this into the surviving $-b/2a$ produces

$$r_3 = \frac{n_\phi^2 \hbar \left(1 - \frac{Z^2\alpha^2}{n_\phi^2}\right)^{1/2}}{mZ\alpha c} \tag{B.6}$$

after applying (B.4) at $n_r = 0$. Since at $n_r = 0$, we have

$$m = m_o \left(1 - \frac{Z^2\alpha^2}{n_\phi^2}\right)^{1/2} \tag{B.7}$$

then

$$r_3 = \frac{n_\phi^2 \hbar}{m_o Z\alpha c} \tag{B.8}$$

This is a different equation for the circular radii to that given by the vortex model (7.23).

# Appendix C

# Computer Code for MVR.exe

A working version of the computer program can be obtained by emailing aleteller@barryispuzzled.com. When implemented, it generates the file MVR.txt and writes data to it. An explanation of the inputs and outputs is given in Appendix D.

```
0  open "MVR.txt" for output as #MVR
'  INPUT DATA
1  print " INPUT (omit quotes in examples below)"
2  input " Enter hydrogen level (e.g. '2P3/2', with upper case orbital letter): ";level$
3  input " Enter proton-electron spin-2 rotation sense (same '+' or opposite '-'): ";sense$
4  input " Enter electron spin-2 major axis rotation angle (degrees): ";rot  ' 119.98 aligns
               ' major axis with momentum field
5  input " Enter Doppler-shift type 'red' or 'blue': ";dopptype$
6  print " writing to MVR.txt ..."
7  print #MVR, " Level: ";level$;" Sense: ";sense$;" Rotation: ";rot; " Doppler shift: ";dopptype$
8  print ""

'  NIST DATA: Hyperfine energy levels (MHz): Horbatsch and Hessels (2016, Tables 3 & 4)
9    n = 0  ' default n if no state in list chosen, for input check line 200
10   select case
11     case level$ = "1S1/2"
12       NIST0 = 3288087922.4160      ' 1S(1/2) ionization energy (F=0)
13       NIST1 = 3288086502.0102      ' 1S(1/2) ionization energy (F=1)
14       n = 1
15       L = 0
16       Sp = 1/2
17     case level$ = "2S1/2"
18       NIST0 = 822025577.0922        ' 2S(1/2) (F=0)
19       NIST1 = 822025399.5354        ' 2S(1/2) (F=1)
20       n = 2
21       L = 0
22       Sp = 1/2
23     case level$ = "2P1/2"
24       NIST0 = 822026546.1359        ' 2P(1/2) (F=0)
25       NIST1 = 822026486.9664        ' 2P(1/2) (F=1)
26       n = 2
```

```
27    L = 1
28    Sp = -1/2
29  case level$ = "2P3/2"
30    NIST0 = 822015547.4967      ' 2P(3/2) (F=1)
31    NIST1 = 822015523.8451      ' 2P(3/2) (F=2)
32    n = 2
33    L = 1
34    Sp = 1/2
35  case level$ = "3S1/2"
36    NIST0 = 365343617.9043      ' 3S(1/2) (F=0)
37    NIST1 = 365343565.2949      ' 3S(1/2) (F=1)
38    n = 3
39    L = 0
40    Sp = 1/2
41  case level$ = "3P1/2"
42    NIST0 = 365343906.4710      ' 3S(1/2) (F=0)
43    NIST1 = 365343888.9392      ' 3S(1/2) (F=1)
44    n = 3
45    L = 1
46    Sp = -1/2
47  case level$ = "3P3/2"
48    NIST0 = 365340647.6119      ' 3P(3/2) (F=1)
49    NIST1 = 365340640.6040      ' 3P(3/2) (F=2)
50    n = 3
51    L = 1
52    Sp = 1/2
53  case level$ = "3D3/2"
54    NIST0 = 365340651.1931      ' 3D(3/2) (F=1)
55    NIST1 = 365340646.9865      ' 3D(3/2) (F=2)
56    n = 3
57    L = 2
58    Sp = -1/2
59  case level$ = "3D5/2"
60    NIST0 = 365339566.8037      ' 3D(5/2) (F=2)
61    NIST1 = 365339564.1003      ' 3D(5/2) (F=3)
62    n = 3
63    L = 2
64    Sp = 1/2
65  case level$ = "4S1/2"
66    NIST0 = 205505309.9525      ' 4S(1/2) (F=0)
67    NIST1 = 205505287.7579      ' 4S(1/2) (F=1)
68    n = 4
69    L = 0
70    Sp = 1/2
71  case level$ = "4P1/2"
72    NIST0 = 205505431.9299      ' 4P(1/2) (F=0)
73    NIST1 = 205505424.5337      ' 4P(1/2) (F=1)
74    n = 4
75    L = 1
76    Sp = -1/2
77  case level$ = "4P3/2"
78    NIST0 = 205504057.1004      ' 4P(3/2) (F=1)
```

```
79    NIST1 = 205504054.1439      ' 4P(3/2) (F=2)
80    n = 4
81    L = 1
82    Sp = 1/2
83  case level$ = "4D3/2"
84    NIST0 = 205504058.6495      ' 4D(3/2) (F=1)
85    NIST1 = 205504056.8748      ' 4D(3/2) (F=2)
86    n = 4
87    L = 2
88    Sp = -1/2
89  case level$ = "4D5/2"
90    NIST0 = 205503601.1722      ' 4D(5/2) (F=2)
91    NIST1 = 205503600.0315      ' 4D(5/2) (F=3)
92    n = 4
93    L = 2
94    Sp = 1/2
95  case level$ = "5S1/2"
96    NIST0 = 131523180.9882      ' 5S(1/2) (F=0)
97    NIST1 = 131523169.6246      ' 5S(1/2) (F=1)
98    n = 5
99    L = 0
100   Sp = 1/2
101  case level$ = "5P1/2"
102   NIST0 = 131523243.5005      ' 5P(1/2) (F=0)
103   NIST1 = 131523239.7136      ' 5P(1/2) (F=1)
104   n = 5
105   L = 1
106   Sp = -1/2
107  case level$ = "5P3/2"
108   NIST0 = 131522539.5886      ' 5P(3/2) (F=1)
109   NIST1 = 131522538.0749      ' 5P(3/2) (F=2)
110   n = 5
111   L = 1
112   Sp = 1/2
113  case level$ = "5D3/2"
114   NIST0 = 131522540.3924      ' 5D(3/2) (F=1)
115   NIST1 = 131522539.4837      ' 5D(3/2) (F=2)
116   n = 5
117   L = 2
118   Sp = -1/2
119  case level$ = "5D5/2"
120   NIST0 = 131522306.1639      ' 5D(5/2) (F=2)
121   NIST1 = 131522305.5800      ' 5D(5/2) (F=3)
122   n = 5
123   L = 2
124   Sp = 1/2
125  case level$ = "6S1/2"
126   NIST0 = 91335431.6017       ' 6S(1/2) (F=0)
127   NIST1 = 91335425.0256       ' 6S(1/2) (F=1)
128   n = 6
129   L = 0
130   Sp = 1/2
```

```
131  case level$ = "6P1/2"
132    NIST0 = 91335467.7972      ' 6P(1/2) (F=0)
133    NIST1 = 91335465.6058      ' 6P(1/2) (F=1)
134    n = 6
135    L = 1
136    Sp = -1/2
137  case level$ = "6P3/2"
138    NIST0 = 91335060.4413      ' 6P(3/2) (F=1)
139    NIST1 = 91335059.5653      ' 6P(3/2) (F=2)
140    n = 6
141    L = 1
142    Sp = 1/2
143  case level$ = "6D3/2"
144    NIST0 = 91335060.9101      ' 6D(3/2) (F=1)
145    NIST1 = 91335060.3843      ' 6D(3/2) (F=2)
146    n = 6
147    L = 2
148    Sp = -1/2
149  case level$ = "6D5/2"
150    NIST0 = 91334925.3612      ' 6D(5/2) (F=2)
151    NIST1 = 91334925.0233      ' 6D(5/2) (F=3)
152    n = 6
153    L = 2
154    Sp = 1/2
155  end select


' INPUT CHECKS
198 if n = 0 then
199 print ""
200  print #MVR, " state not available: run again"
201  close #MVR
202  end
203 end if

204 if sense$ = "+" or sense$ = "-" then
205 else
206  print #MVR, " enter + or -: run again"
207  close #MVR
208  end
209 end if

210 if rot<0 or rot>360 then
211  print #MVR, " enter angle between 0 and 360"
212  close #MVR
213  end
214 end if

215 if dopptype$ = "red" or dopptype$ = "blue" then
216 else
217  print #MVR, " enter 'red' or 'blue'"
218  close #MVR
```

```
219 end
220 end if

' HYPERFINE MULTIPLIER RULE
' correction rule is 0.0000002159927198*hypconst/n
221 select case
222  case L=0 and Sp=1/2  ' nS1/2
223   hypconst = 1
224   hypconst1$ = "1"
225  case L=1 and Sp=-1/2  ' nP1/2
226   hypconst = 1/3
227   hypconst1$ = "1/3"
228  case L=1 and Sp=1/2  ' nP3/2
229   hypconst = 2/15
230   hypconst1$ = "2/15"
231  case L=2 and Sp=-1/2  ' nD3/2
232   hypconst = 2/25
233   hypconst1$ = "2/25"
234  case L=2 and Sp=1/2  ' nD5/2
235   hypconst = 50/972
236   hypconst1$ = "50/972"
237 end select


' CODATA (June 2019)
238  alpha1 = 0.0072973525693   ' NIST: fine structure constant
239  alpha2 = alpha1*alpha1     ' square of alpha1
240  Rydfreq = 3289841960.2508  ' (MHz) NIST: Rydberg frequency, (1/2)*mo*alpha2*clight^2*10^(-6)/h
241  Me = 9.1093837015*10^(-31)  ' (kg) NIST: electron rest mass
242  clight = 299792458         ' (m/s), speed of light, CODATA (2018, 126)
243  M = 0.000544617021488      ' NIST: e- to p+ rest mass ratio
244  pi = 3.141592653589793
245  h = 6.62607015*10^(-34)    '(J/Hz) Planck constant, CODATA (2018, 142)

' SOMMERFELD/MVR FINE STRUCTURE QUANTUM NUMBERS
246  na = L + Sp + 0.5      ' azimuthal quantum number
247  nr = n - na           ' radial quantum number
248  Y = nr + sqr(na*na - Kalpha^2*alpha2)
249  Y2 = Y*Y

' OUTPUT
250 Print #MVR, " Quantum numbers -> QM: n = "; n; "  L = "; L; "  Spin = ";Sp; ";  MVR: na = ";na;"  nr = ";nr
251 Print #MVR, ""

' OUTPUT TEMPLATES
252  template1$ = "##########.####"
253  template2$ = "#####.##########"
254  template3$ = "##.###############"

' CALCULATE SOMMERFELD F-S ENERGY
255 print #MVR, " FINE STRUCTURE ENERGIES (MHz, Kalpha = 1)"
256 gosub [Sommerfeldfs]  ' Kalpha=1, no change to f-s constant
257 print #MVR, " (1) Sommerfeld f-s  = ";using (template1$, SommEnergy)
```

' CALCULATE MVR F-S ENERGY WITHOUT FINE STRUCTURE CONSTANT ADJUSTMENT

258  Kalpha = 1    ' prepare for MVRfs calc

259  gosub [MVRfs]  ' returns erad

260  VortexEnergy2 = erad  ' MVR fine structure energy at Kalpha = 1

261  print #MVR, " (2) MVR f-s    = ";using (template1$, VortexEnergy2)


' MEMORY STORAGE

262  nastore = na  ' store input quantum numbers for later reset

263  nrstore = nr

264  na = 1      ' initialize 1S1/2 state quantum numbers

265  nr = 0


' KALPHA MULTIPLIER FOR F-S CONSTANT, SEARCHES: 1ST PASS 1S1/2, 2ND PASS IS REQD LEVEL
' USING 1S1/2 EMISSION COORDS

300  for Kloop = 1 to 2  '

301  Kinc = 0        ' initialize Kalpha search increment

302  Kalphastore = 0.8 ' initialize Kalpha search value

303  for Kacc = 1 to 16  ' locates decimal place searched

304  for Kdigit = 0 to 10  ' number inserted in decimal place

305  Kinc = Kdigit/10^Kacc

306  Kalpha = Kalphastore + Kinc

307  gosub [MVRfs]  ' using input quantum numbers and Y

308  VortexEnergy1 = erad  ' MVR fine structure energy at trial Kalpha

' switch to 1S1/2 circle quantum numbers for emission coordinates

309  Y = nr + sqr(na^2 - Kalpha^2*alpha2) ' uses lines 246/247 1st pass, 343/344 2nd pass

310  Y2 = Y*Y

311  if Kloop = 1 then ' only need to calc coordinates (1S1/2) on 1st pass

312  theta = 0  ' arbitrary value, not really relevant

' Confine location of reqd SAM ring to 1S1/2 OAM ring & ask what coords (Xstore, gamma) produce P.E. = K.E.

313  gosub [coordinates]  ' return coordinates x = Xstore and y = gamma of point on electron ring (circ/per/apog)

314  Xstore1 = Xstore  ' store 1S1/2 coords for next pass Doppler calc

315  gamma1 = gamma

316  end if

317  gosub [Doppler]   ' calculate Doppler blue-shift due to proton motion towards 1S1/2 ring location

318  if Kloop = 1 then  ' On 1st pass, find Kalpha that places 1S1/2 f-s energy at mid-point of hyperfine levels
       ' (pre-Doppler)

319  Ksearch = VortexEnergy1 - ((3288087922.4160 + 3288086502.0102)/(2*Doppler1))

320  else        ' On 2nd pass, find Kalpha that places reqd f-s energy at mid-point of hyperfine levels (pre-Doppler)

321  Ksearch = VortexEnergy1 - ((NIST0 + NIST1)/(2*Doppler1))

322  end if

' change of sign detection in line 319 or 321

323  if Ksearch < 0 then  ' mark the sign

324  Kmark = 0

325  else

326  Kmark = 1

327  end if

328  if Kdigit > 0 then ' only perform a sign comparison after 1st Kdigit pass

329  if Kmark <> Kstoremark then   ' sign change in Ksearch

330  Kalpha = Kalphastore + (Kdigit - 1)/10^Kacc ' return to previous digit in decimal place

331  Kalphastore = Kalpha

332  if Kmark = 1 then  ' reset sign-change marker Kmark to previous value

333  Kmark = 0

```
334  else
335    Kmark = 1
336    end if
337    Kdigit = 10  ' leave the Kdigit loop
338  end if
339  end if
340  Kstoremark = Kmark
341  next Kdigit
342  next Kacc
343  na = nastore  ' reset to reqd level quantum numbers for 2nd Kloop pass
344  nr = nrstore
345  Xstore = Xstore1  ' use 1S1/2 emission coords on 2nd pass
346  gamma = gamma1
347  Kalphastore1 = Kalpha  ' store Kalpha for reqd level printout on exiting Kloop
348  next Kloop


'  COORDINATES OF REQUIRED LEVEL PERIGEE/APOGEE/CIRCLE (KALPHA = 1)
349  theta = 0
350  Kalpha = 1
351  Y = nr + sqr(na*na - Kalpha^2*alpha2)  ' based on 2nd pass (na, nr) in lines 343/344
352  Y2 = Y*Y
353 print #MVR, ""
354 print #MVR, " COORDINATES (multiplied by ground state radius)"
355 if L = (n - 1) and Sp = 1/2 then
356   print #MVR, " Circle (Kalpha = 1)"
357 else
358   print #MVR, " Perigee (Kalpha = 1)"
359 end if
360 gosub [coordinates]   ' input Kalpha = 1, returns Xstore and gamma for perigee (where P.E.=K.E.)
361 gosub [printcoords]
362 if L = (n - 1) and Sp = 1/2 then
363 else
364   print #MVR, " Apogee (Kalpha = 1)"
365   theta = pi
366   Kalpha = 1
367   gosub [coordinates]   ' input Kalpha = 1, returns Xstore and gamma for apogee (where (P.E.=K.E.)
368   gosub [printcoords]
369 end if
370  n = na + nr   ' based on 2nd pass (na, nr) in lines 343/344
371  Kalpha = Kalphastore1  ' reset Kalpha for reqd level at hyperfine mid-point
372  Xstore = Xstore1   ' 1S1/2 x coord
373  gamma = gamma1     ' 1S1/2 circle radius
374  NISTavD = (NIST0 + NIST1)/(2*Doppler1)
375  NISTav = (NIST0 + NIST1)/2
376  deviation = (NIST0 - NISTav)/(NISTav)
377  gosub [printout1]


'  SEARCH FOR ELECTRON SPIN-2 ECCENTRICITY FOR ALIGNEMENT WITH HIGHER H-F VALUE
399  searchpass = 0  ' 0 is pass, 1 is fail
400  Hinc = 0
401  Hstore = 0  ' initialize search value
```

```
402 for Hacc = 1 to 16  ' locates decimal place searched, was 16
403 for Hdigit = 0 to 10  ' number in decimal place
404 Hinc = Hdigit/10^Hacc
405 e = Hstore + Hinc
406 hypercalcstore = hypercalc  ' stored for [printout2]
407 if e > 0.999999999999 then
408   e = 0.9999999999  ' prevents sqr negative in k2 from [Hfmodel2], simulates e = 1.0
409 end if
410 gosub [Hfmodel2]
411 hypercalc1 = abs(hypercalc)
412 Hsearch = hypercalc1 - deviation  ' condition for MVR calc approaching higher NIST value
' change of sign detection in line 413
413 if Hsearch < 0 then
414   Hmark = 0
415 else
416   Hmark = 1
417 end if
418 if Hdigit > 0 then  ' sign comparison only after 1st Hdigit pass
419   if Hmark <> Hstoremark then  ' check for sign change of Fsearch
420     e = Hstore + (Hdigit - 1)/10^Hacc  ' return to previous digit in decimal place
421     Hstore = e
422     if Hmark = 1 then  ' reset sign-change marker Kmark to previous value as decimal place reverts
423       Hmark = 0
424     else
425       Hmark = 1
426     end if
427     Hdigit = 10  ' change of sign then leave Hdigit loop
428   end if
429 end if
430 if Hdigit = 10 and Hsearch > 0 then  ' if still no sign change with excess digit then leave
431   Hacc = 16  ' leave Hacc loop
432   searchpass = 1  ' fail search, for [printout2]
433 end if
434 Hstoremark = Hmark
435 next Hdigit
436 next Hacc
437 gosub [printout2]
438 print #MVR, ""
439 close #MVR
440 end


' HYPERFINE CORRECTION FACTOR THAT MULTIPLIES VortexEnergy1*Doppler1
' INPUT: e, n, pi, Kalpha, M, Xstore, alpha1, Y, rot
' OUTPUT: perim, hypercalc
[Hfmodel2]
' Cayley is quick but inaccurate at high e (9dp e = 0.92)
' Maclaurin in slow but accurate at high e (16dp n = 125 at e = 0.92)
' Cayley perimeter is P = 4*a*integral[0-pi/2]*sqr(1 - k^2*cos^2(th))*d(th) (Chandrupatla and Osler, 2010, p.123)

500 if e > 0.9 then  ' Cayley series convergence to 8d.p.: e = 0.93, k2^16, e = 0.99, k2^8
' correct: to 4dp e = 0.7; 5dp e = 0.8; 6dp e = 0.85; 8dp e = 0.9; 9dp e = 0.92
```

```
501  k2 = sqr(1 - e*e)
502  perima = 1 + 0.5*k2^2*(log(4/k2) - (1/2))
503  perimb = (3*k2^4/16)*(log(4/k2) - 1 - (1/12))
504  perimc = (45*k2^6/384)*(log(4/k2) - 1 - (1/6) - (1/30))
505  perimd = (1575*k2^8/18432)*(log(4/k2) - 1 - (1/6) - (1/15) - (1/56))
506  perime = (99225*k2^10/1474560)*(log(4/k2) - 1 - (1/6) - (1/15) - (1/28) -(1/90))
507  perimf = (9823275*k2^12/176947200)*(log(4/k2) - 1 - (1/6) - (1/15) - (1/28) -(1/45) -(1/132))
508  perimg = (1404728325*k2^14/29727129600)*(log(4/k2) - 1 - (1/6) - (1/15) - (1/28) -(1/45) -(1/66) - (1/182))
509  perimh = (273922023375*k2^16/6658877030400)*(log(4/k2) - 1 - (1/6) - (1/15) - (1/28) -(1/45) -(1/66)
           - (1/91) - (1/240))
510  perimi = (69850115960625*k2^18/1917756584755200)*(log(4/k2) - 1 - (1/6) - (1/15) - (1/28) -(1/45)
           -(1/66) - (1/91) -(1/120) - (1/306))
511  perim = (perima + perimb + perimc + perimd + perime + perimf + perimg + perimh + perimi)/(pi/2)
512 else   ' MacLaurin expansion for perimeter
' convergence 16dp: e = 0.7 plimit = 34; e = 0.8 plimit = 54; e = 0.85 plimit = 68; e = 0.9 plimit = 101;
'e = 0.92 plimit = 125
513   perim = 1  ' initialization, perimeter = 2*pi*a*perim, with a = r2/perim for const perimeter
514   plimit = 102
515  for nl = 1 to plimit
516   tstore = 1
517   for nn = 1 to nl
518     tcalc = 2*nn - 1
519     tstore = tstore*tcalc
520   next nn
521   pstore = 1
522   for mm = 1 to nl
523     pcalc = 2*mm
524     pstore = pstore*pcalc
525   next mm
526   coeff0 = 1/(2*nl - 1)
527   coeff3 = coeff0*(tstore/pstore)^2
528   term = (-1)*coeff3*e^(2*nl)
529   perim = perim + term
530   if abs(term) < 0.00000000000001 then
531    nl = plimit   ' leave nl loop
532   end if
533  next nl
534 end if


' calculate line integral of field momentum around electron spin-2 circuit (only azimuthal component, no radial)
535  theta = 0
536  totint = 0
537  factor = 0
538  thlimit = 100    ' convergence to 10dp for fieldmom, to 15dp for hypercalc (tested against thlimit = 400000)
539  for thloop = 1 to thlimit
540   theta = theta + 2*pi/thlimit
' perimeter = 2*pi*a*perim, a = r2/perim, perimeter kept constant
541   r = (1/perim)*(1 - e*e)*Kalpha*alpha1/sqr(1 - Kalpha*Kalpha*alpha1^2) ' (9.6)
542   rad = r/(1 - e*cos(theta))                ' (9.6)
' reciprocal field radius
543   T01 = Xstore*Xstore + (1 - M)*(1 - M)
544   T02 = Xstore*cos(theta + (rot*pi/180))
```

545   T03 = (1 - M)*sin(theta + (rot*pi/180))
546   T04 = 1 + (rad^2/T01) + (2*rad*(T02 + T03)/T01)
547   T05 = (1/T01)*(1/T04)*cos(theta + (rot*pi/180))
548   T06 = (-2)*(Xstore + rad*cos(theta + (rot*pi/180)))
549   T07 = (T06/(T01^2))*(1/T04)^2*(T02 + T03)
550   factor = (T05 + T07)
551   totint = totint + factor
552   next thloop


553   fieldmom = totint*2*pi/thlimit
554   if sense$ = "+" then    ' switch sign if same sense spin-2
555   fieldmom = (-1)*fieldmom
556   end if
557   hypercalc1 = sqr(1 + alpha2*Kalpha^2/(Y*Y))*sqr(1 - (alpha2*Kalpha^2/(2*(Y*Y + alpha2*Kalpha^2))))
558   hypercalc = sqr(2)*fieldmom*hypercalc1*(alpha2*Kalpha^2/(1 - alpha2*Kalpha^2))   ' MVR h-f correction
559   return


' SOMMERFELD-DIRAC FINE STRUCTURE ENERGY (calculated at Kalpha = 1)
' INPUT: na, nr, alpha1, alpha2, M, Rydfreq
' OUTPUT: DiracEnergy
[Sommerfeldfs]
570   redmass1 = 1/(1 + M)    ' reduced mass based on rest masses
571   Somm01 = na^2 - alpha2
572   Somm02 = sqr(Somm01)
573   Somm03 = nr   ' Sommerfeld's radial quantum number
574   Somm04 = alpha1/(Somm02 + Somm03)
575   Somm05 = 1 + Somm04^2
576   Somm06 = 1/(sqr(Somm05))
578   Somm07 = 1 - Somm06
579   SommEnergy = Somm07*redmass1*Rydfreq*2/alpha2
580   return


' MVR FINE STRUCTURE ENERGY (calculate with Kalpha < 1 f-s adjustment)
' INPUT: na, nr, Kalpha, alpha2, M, Rydfreq
' OUTPUT: erad
[MVRfs]
581   vortex1 = nr + sqr(na*na - Kalpha^2*alpha2)
582   vortex2 = vortex1*vortex1
583   vortex3 = Kalpha^2*alpha2/(vortex2)
584   vortex4 = vortex3/(2*(1 + vortex3))   ' factor of 2 from (7.17)
585   vortex5 = sqr(1 - vortex4)
586   vortex6 = vortex2 + Kalpha^2*alpha2
587   k1 = M/vortex5
588   redmass2 = 1/(1 + k1)   ' reduced mass with relativistic electron mass
589   erad = redmass2*Kalpha^2*Rydfreq/(vortex5*vortex6)   ' energy of level Rydfreq = 0.5*alpha^2*clight^2
            ' *Kalpha^2
590   return


' PROTON-ELECTRON SEPARATION WHERE RING ENERGY EQUALS INTRUDING PROTON
  FIELD ENERGY (P.E. = K.E.)
[coordinates]
' INPUT: na, nr, Y, Y2, alpha1, alpha2, theta, Kalpha, M

```
' OUTPUT: Xstore (x coord), gamma (ring radius)
600 if nr = 0 then     ' negative error avoidance in sqr
601  ecc = 0
602 else
603  ecc = sqr(1 - (na*na - Kalpha^2*alpha2)/Y2)  ' eccentricity of ellipse
604 end if


' EQUATION OF ELLIPSE AS FUNCTION OF ANGLE THETA
605 g1 = 1 + ecc*cos(theta)          ' theta = 0 (min, perigee) or theta = pi (max, apogee)
606 g2 = Y2*(1 - ecc*ecc)
607 gamma = g2/(g1*(1 - Kalpha^2*alpha2))  ' ring radius, as fraction of 1S1/2 state radius


' SEARCH ROUTINE FOR LOCATION OF EMISSION, X COORDINATE RELATIVE TO
'PROTON SPIN-2 CENTRE
608 Xinc = 0          ' initialize increment for X search
609 Xstore = 0          ' initialize memorized value of X
610 for acc = 0 to 14     ' variable controlling decimal place searched
611 for msearch = 0 to 150 ' raised upper limit for high x, only 0-10 reqd for decimal places

612 Xinc = msearch/10^(acc) ' increment added to previous trial X
613 X = Xstore + Xinc     ' stores trial value of X coordinate, as fraction of ground state radius
614 XX = X/gamma          ' first equation (7.54) (The Vortex Atom: A New Paradigm)
615 R1 = M/gamma          ' second equation (7.54)
616 R2 = Kalpha*alpha1/(gamma*sqr(1 - Kalpha^2*alpha2))  ' third equation (7.54)


' CALCULATION OF F (7.51)

617 FX1 = sqr(1 - (Kalpha^2*alpha2/(na*na)))
618 FX2 = 1 - Kalpha^2*alpha2
619 FX3 = FX1/FX2


' INTEGRAL (7.55)
620 Fint1 = 1 - R1
621 Fint2 = 1 + (XX*XX + R2*R2)/(Fint1*Fint1)
622 Fint3 = 1/(Fint1*sqr(Fint2))
623 Fint4 = R2*R2*(XX*XX + Fint1*Fint1)/(Fint1^4*Fint2^2)
624 Fint5 = Fint4*Fint4
625 Fint = Fint3*(1 + (3/4)*Fint3^4*Fint4 + (105/64)*Fint3^8*Fint5)
626 FX = FX3*Fint/gamma


' EQUALITY (7.56), AT 1S1/2 OSCILLATION BOUNDARY, AIM IS MINIMIZE Fsearch = 0
627 Fsearch = FX - 0.5/(Y2 + Kalpha^2*alpha2)  ' resident elect ring energy equals prot field energy
628 if Fsearch < 0 then  ' attach marker value to sign of Fsearch
629  mark = 0
630 else
631  mark = 1
632 end if


633 if msearch > 0 then  ' only compare Fsearch sign with previous one only after first pass
634  if mark <> storemark then
635   Xstore = Xstore + (msearch - 1)/10^acc  ' return to previous digit in search
636   msearch = 10
```

637   end if
638   end if

639   storemark = mark   ' store present Fsearch sign


640   next msearch
641   next acc
642   return


[Doppler]
' DOPPLER SHIFT BASED ON 1S1/2 RING COORDINATES (X, Y) (COSINE CALC)
' INPUT: Xstore, gamma, Kalpha, M, alpha1, alpha2, Y2, coz1. dopptype$
' OUTPUT: Doppler1
' Assume emission is away from proton along its spin-2 radial direction (i.e. at 29.9842 degrees to + x axis)
700   dopplera = Kalpha*alpha1/(sqr(2)*sqr(Y2 + Kalpha^2*alpha2))   ' electron speed in its ring
701   dopplerb = M/sqr(1 - (alpha2/(2*(Y2 + Kalpha^2*alpha2))))   ' translation to proton speed towards electron
702   coz1 = Xstore/sqr(Xstore*Xstore + (gamma - M)^2)   ' based on emission n-circle quantum numbers
' A negative sign after the first ' 1 ' produces a Doppler red shift
703   if dopptype$ = "blue" then
704   Doppler1 = (1 + dopplera*dopplerb*coz1)/sqr(1 - (dopplera*dopplerb*coz1)^2)   ' blue shift,
            ' proton advances towards emission
705   else
706   Doppler1 = (1 - dopplera*dopplerb*coz1)/sqr(1 - (dopplera*dopplerb*coz1)^2)   ' red shift, proton moves
            ' away from emission
707   end if
708   return


[printcoords]
' INPUT: Xstore, gamma, M
' OUTPUT: print
709 if Kalpha = 1 then
710   print #MVR, " (3) X = ";using (template2$, Xstore);" (p-e ring separation)"
711   print #MVR, " (4) Y = ";using (template2$, gamma);" (ring radius)"
712   print #MVR, " (5) X*X + Y*Y = ";using (template2$, (Xstore^2 + (gamma - M)^2));" (radius^2 from proton)"
713 else
714   print #MVR, " (6) X = ";using (template2$, Xstore);" (p-e ring separation)"
715   print #MVR, " (7) Y = ";using (template2$, gamma);" (ring radius)"
716   print #MVR, " (8) X*X + Y*Y = ";using (template2$, (Xstore^2 + (gamma - M)^2));" (radius^2 from proton)"
717 end if
718 return
[printout1]
' INPUT: coz1, Doppler1, NIST1, NIST0, VortexEnergy1, NISTavD, NISTav
' OUTPUT: print
800   print #MVR, " Collapse to 1S1/2 radius (adjusted Kalpha)"
801   gosub [printcoords]
802   print #MVR, ""
803   print #MVR, " DOPPLER VALUES (at 1S1/2 radius)"
804   print #MVR, " (9)  cosine    = ";using (template3$, coz1); " (cosine of emission angle)"
805   print #MVR, " (10) Dopp factor = ";using (template3$, (Doppler1));
806 if dopptype$ = "blue" then
807   print #MVR, " (blue shifted emission)"
808 else

809 print #MVR, " (red shifted emission)"
810 end if
811 print #MVR, ""
812 print #MVR, " PRE-EMISSION (before Doppler effect)"

813 print #MVR, " (11) low h-f/Dopp    = ";using (template1$,NIST1/Doppler1);" high h-f/Dopp   = ";
    using (template1$,NIST0/Doppler1)
814 print #MVR, " (12) MVR h-f mid-pt = ";using (template1$, VortexEnergy1); "
    (Kalpha: alignment h-f energies mid-pt)"
815 print #MVR, " (13) expt h-f mid-pt = ";using (template1$,NISTavD); "  (pre-Doppler h-f energies mid-point)"
816 print #MVR, ""
817 print #MVR, " POST-EMISSION (after Doppler effect)"
818 print #MVR, " (14) MVRDopp        = ";using (template1$, (VortexEnergy1*Doppler1))
819 print #MVR, " (15) expt h-f mid-pt = ";using (template1$,NISTav)
820 return


[printout2]
' INPUT: perim, hyconst, n, hyconst1$, searchpass, hypercalcstore, deviation, Doppler1, hypercalc, NIST1, NIST0,
' Kalpha, e
' OUTPUT: print
900 print #MVR, ""
901 print #MVR, " HYPERFINE ENERGIES"
902 if searchpass = 0 then
903  print #MVR, " (16) spin-2 major axis = "; using (template3$,1/perim);" (x spin-2 circle rad. same perim.)"
904 else
905  print #MVR, " (16) spin-2 major axis =   no value possible at this rotation angle"
906 end if
' The value of the magnitude that follows is obtained from (9.22)
907 print #MVR, " (17) h-f rule correction magnitude  = ";using (template3$,0.000000215911341*hypconst/Y);"
    (mult. const = ";hypconst1$;")"
908 if searchpass = 0 then
909  print #MVR, " (18) MVR proton field correction   = ";using (template3$,hypercalcstore)
910 else
911  print #MVR, " (18) MVR proton field correction   = no value possible at this rotation angle"
912 end if
913 print #MVR, " (19) expt correction magnitude     = ";using (template3$, deviation)
914 if sense$ = "+" then
915 print #MVR, " (20)   MVR low  = ";using (template1$, (VortexEnergy1*Doppler1));"
    (MVR f-s without correction)"
916 print #MVR, " (21)   MVR low  = ";using (template1$, (VortexEnergy1*Doppler1)*
    (1 - 0.000000215911341*hypconst/Y));" (MVR with rule)"
917 if searchpass = 0 then   ' only print h-f correction if search was successful
918  print #MVR, " (22) -> MVR low  = ";using (template1$, (VortexEnergy1*Doppler1*(1 + hypercalcstore)));"
    (MVR spin-2 proton field) <-"
919 else
920  print #MVR, " (22) -> No value from MVR spin-2 proton field"
921 end if
922  print #MVR, " (23) -> expt low = ";using (template1$, NIST1);" (Horbatsch and Hessels, 2016) <-"
923 else
924 print #MVR, " (20)   MVR high  = ";using (template1$, (VortexEnergy1*Doppler1));"
    (MVR f-s without correction)"
925 print #MVR, " (21)   MVR high = ";using (template1$, (VortexEnergy1*Doppler1)*

```
         (1 + 0.000000215911341*hypconst/Y));" (MVR with rule)"
926  if searchpass = 0 then  ' only print h-f correction if search was successful
927    print #MVR, " (22) -> MVR high  = ";using (template1$, (VortexEnergy1*Doppler1*(1 + hypercalcstore)));
         " (MVR spin-2 proton field) <-"
928  else
929    print #MVR, " (22) -> No value from MVR spin-2 proton field"
930  end if
931    print #MVR, " (23) -> expt high = "; using (template1$,NIST0);" (Horbatsch and Hessels, 2016) <-"
932  end if  ' end sense$
933  print #MVR, ""
934  print #MVR, " PARAMETERS SEARCHED FOR ALIGNMENT"
935  print #MVR, " (24) -> Kalpha f-s multiplier = ";using (template3$, Kalpha);" <-"
936  if searchpass = 0 then
937    print #MVR, " (25) -> spin-2 eccentricity  = ";using (template3$,e);" <-"
938  else
939    print #MVR, " (25) -> spin-2 eccentricity  =  no value possible at this rotation angle <-"
940  end if
941  print #MVR, ""
942  print #MVR, " *** Free license, Barry R. Clarke (aleteller@barryispuzzled.com), 11 July 2020 ***"
943  return


'  *** END OF PROGRAM ***
```

# Appendix D

# Computer Program Input and Output

The program MVR.exe was originally written in Liberty BASIC and compiled into executable code using an application developed by R. T. Russell. It writes its output to the text file MVR.txt. All line numbers (e.g., line 607) refer to the listed program in Appendix C. What follows is a description of all the input and output values to the program.

## D.1 User Keyed Input to Computer Program MVR.exe

All inputs have the quotation marks omitted.

(1) The level under calculation. For example, '2P3/2'.
(2) Selection of one of the two hyperfine levels for the state chosen in (1). The '+' means that the proton and electron spin-2 rotations are same sense, while the '−' means opposite sense. With the proton spin-2 field advancing toward the electron, the former gives the lower hyperfine frequency while the latter gives the upper.
(3) The spin-2 ellipse rotation angle in degrees. The major axis is at zero degrees when along the positive $x$ axis, where the left focus is at the origin. The angle is then taken counterclockwise. A value of 135° (to the nearest degree) gives the minimum eccentricity required for alignment for all 48 hyperfine levels in the program with the experimental values, except the level $4D_{5/2}$

which requires an input angle 127°, $5D_{5/2}$ which works at 125°, and $6D_{5/2}$ which gives results at 124°.

(4) The Doppler shift type: 'red' or 'blue'. The proton is advancing toward the stationary electron ring which is at the oscillation boundary on emitting. A red shift sends the emission through the proton so that on emergence the proton recedes from it. A blue shift has the emission running away from the proton which is approaching it.

## D.2  Output from Computer Program MVR.exe

(1) The Sommerfeld fine-structure frequency (MHz) for a given level calculated with the CODATA fine-structure constant (Mohr *et al.* 2018). This is obtained from lines 570–580 of the program.

(2) The Mass Vortex Ring (MVR) theory frequency (MHz) for a given level, lines 581–590. Again this is based on the CODATA fine-structure constant.

(3) For the chosen level, using the CODATA fine-structure constant (so *Kalpha* = 1), this is the separation of the proton and electron rings along the $x$ axis expressed as a fraction of the ground state spin-3 radius $r_{gs}$. For an elliptic ring, there is an entry for the perigee and apogee. This is calculated at the oscillation boundary to satisfy the equality of the momentum field and resident electron energy (7.56); see line 627.

(4) For the chosen level, this is the spin-3 ring radius as a fraction of the ground state spin-3 radius $r_{gs}$, using *Kalpha* = 1. For an elliptic ring, there is an entry for the perigee and apogee; see line 607.

(5) For the chosen level, this is the square of the length of line connecting the spin-2 centers of the proton and electron, using (3) and (4). However, the spin-3 radius of the proton has been subtracted from $Y$ for this calculation; see line 712.

(6) This is calculated on the $1S_{1/2}$ ring (to which all levels collapse prior to absorption or emission), with *Kalpha* adjusted to align the MVR fine-structure frequency with the experimental

hyperfine mid-point frequency. It is the separation of the proton and electron rings along the $x$ axis expressed as a fraction of the ground state spin-3 radius $r_{gs}$.

(7) This is calculated on the $1S_{1/2}$ ring, with *Kalpha* adjusted to align the MVR fine-structure frequency with the experimental hyperfine mid-point frequency. It is the spin-3 ring radius as a fraction of the ground state spin-3 radius $r_{gs}$.

(8) This is the square of the length of line connecting the spin-2 centers of the proton and electron, using (6) and (7). However, the spin-3 radius of the proton has been subtracted from $Y$ for this calculation; see line 712.

(9) Using the optimum *Kalpha* for the Doppler shift calculation, this is the fraction of (6) divided by the square root of (8); see Figure 8.3 and line 702.

(10) Using the optimum *Kalpha* for the Doppler shift calculation, this is the blue-shift multiplier on emission from the ground state ring; see lines 704 and 706.

(11) These are hypothesized frequencies, calculated by dividing the experimental low and high hyperfine frequencies by the Doppler multiplier (10). This gives a pre-Doppler (or pre-emission) mid-point value; see line 813.

(12) This is the MVR fine-structure frequency, aligned with the mid-point of the experimental low and high hyperfine frequencies through a *Kalpha* search, before applying the Doppler shift; see line 814.

(13) This is a hypothesized experimental fine-structure frequency, calculated by dividing the mid-point of the experimental low and high hyperfine frequencies by the Doppler multiplier (10). This gives a pre-Doppler (or pre-emission) mid-point value; see line 815.

(14) This is the MVR fine-structure frequency aligned with the mid-point of the experimental low and high hyperfine frequencies through a *Kalpha* search, after applying the Doppler shift; see line 818.

(15) The mid-point of the experimental low and high hyperfine frequencies; see line 819.

(16) Using the electron spin-2 radius, the perimeter of the spin-2 circuit is calculated. When this circuit is deformed into an ellipse, this perimeter is retained. This gives the major axis of the spin-2 ellipse as a multiplier of the circular spin-2 radius (circle = 1); see lines 500–534, 903.

(17) The hyperfine rule (9.22), which is independent of the eccentricity search calculation; see line 907.

(18) The difference between the MVR (*Kalpha* searched) mid-point of the hyperfine frequencies and the low or high hyperfine frequency. This is obtained from a search of the electron spin-2 eccentricity (to align with the experimental value) which varies the azimuthal proton field momentum affecting the circuit; see line 909.

(19) The difference between the experimental mid-point of the hyperfine frequencies and the low or high hyperfine frequency; see line 913.

(20) Repetition of (14).

(21) The MVR low or high hyperfine frequency calculated from the *Kalpha*-searched hyperfine mid-point frequency and the hyperfine rule (17); see line 916 or 925.

(22) The MVR low or high hyperfine frequency calculated from the *Kalpha*-searched hyperfine mid-point frequency and the eccentricity-searched (18); see line 918 or 927.

(23) The experimental low or high hyperfine frequency; see line 931.

(24) The fine-structure multiplier *Kalpha* that produces alignment of the MVR fine-structure frequency with the experimental mid-point of the two hyperfine frequencies for a given level; see line 935.

(25) The spin-2 eccentricity that brings the experimental mid-point of the two hyperfine frequencies into coincidence with the low or high hyperfine frequency; see line 937.

## D.3   Output Examples

### D.3.1   *Example 1: $1S_{1/2}$ low hyperfine, red shift*

Level: 1S1/2  Sense: +  Rotation: 135  Doppler shift: red
Quantum numbers -> QM: n = 1  L = 0  Spin = 0.5; MVR: na = 1  nr = 0

FINE STRUCTURE ENERGIES (MHz, Kalpha = 1)
(1) Sommerfeld f-s  =  3288095006.0435
(2) MVR f-s      =  3288094981.9247

COORDINATES (multiplied by ground state radius)
Circle (Kalpha = 1)
(3) X =   1.7324126795 (p-e ring separation)
(4) Y =   1.0000000000 (ring radius)
(5) X*X + Y*Y =   4.0001647545 (radius^2 from proton)
Collapse to 1S1/2 radius (adjusted Kalpha)
(6) X =   1.7324126795 (p-e ring separation)
(7) Y =   1.0000000000 (ring radius)
(8) X*X + Y*Y =   4.0001647546 (radius^2 from proton)

DOPPLER VALUES (at 1S1/2 radius)
(9) cosine  = 0.8661885013563859 (cosine of emission angle)
(10) Dopp factor = 0.9999975657833692 (red shifted emission)

PRE-EMISSION (before Doppler effect)
(11) low h-f/Dopp = 3288094505.9445  high h-f/Dopp  = 3288095926.3538
(12) MVR h-f mid-pt = 3288095216.1492  (Kalpha: alignment h-f energies mid-pt)
(13) expt h-f mid-pt = 3288095216.1492  (pre-Doppler h-f energies mid-point)

POST-EMISSION (after Doppler effect)
(14) MVRDopp = 3288087212.2131
(15) expt h-f mid-pt = 3288087212.2131

HYPERFINE ENERGIES
(16) spin-2 major axis = 1.3576307945779310 (x spin-2 circle rad. same perim.)
(17) h-f rule correction magnitude = 0.0000002159170900 (mult. const = 1)
(18) MVR proton field correction  = -0.0000002159927198
(19) expt correction magnitude  = 0.0000002159927198
(20)   MVR low = 3288087212.2131 (MVR f-s without correction)
(21)   MVR low = 3288086502.2589 (MVR with rule)
(22) -> MVR low = 3288086502.0102 (MVR spin-2 proton field) <-
(23) -> expt low = 3288086502.0102 (Horbatsch and Hessels, 2016) <-

PARAMETERS SEARCHED FOR ALIGNMENT

(24) -> Kalpha f-s multiplier =  1.0000000356165786 <-

(25) -> spin-2 eccentricity  = 0.9116435022877840 <-

## D.3.2  *Example 2: $1S_{1/2}$ high hyperfine, red shift*

Level: 1S1/2  Sense: -  Rotation: 135  Doppler shift: red

Quantum numbers -> QM: n = 1  L = 0  Spin = 0.5; MVR: na = 1  nr = 0

FINE STRUCTURE ENERGIES (MHz, Kalpha = 1)

(1) Sommerfeld f-s = 3288095006.0435

(2) MVR f-s     = 3288094981.9247

COORDINATES (multiplied by ground state radius)

Circle (Kalpha = 1)

(3) X =   1.7324126795 (p-e ring separation)

(4) Y =   1.0000000000 (ring radius)

(5) X*X + Y*Y =   4.0001647545 (radius^2 from proton)

Collapse to 1S1/2 radius (adjusted Kalpha)

(6) X =   1.7324126795 (p-e ring separation)

(7) Y =   1.0000000000 (ring radius)

(8) X*X + Y*Y =   4.0001647546 (radius^2 from proton)

DOPPLER VALUES (at 1S1/2 radius)

(9)  cosine = 0.8661885013563859 (cosine of emission angle)

(10) Dopp factor = 0.9999975657833692 (red shifted emission)

PRE-EMISSION (before Doppler effect)

(11) low h-f/Dopp = 3288094505.9445  high h-f/Dopp  = 3288095926.3538

(12) MVR h-f mid-pt = 3288095216.1492  (Kalpha: alignment h-f energies mid-pt)

(13) expt h-f mid-pt = 3288095216.1492 (pre-Doppler h-f energies mid-point)

POST-EMISSION (after Doppler effect)

(14) MVRDopp = 3288087212.2131

(15) expt h-f mid-pt = 3288087212.2131

HYPERFINE ENERGIES

(16) spin-2 major axis = 1.3576307945779310 (x spin-2 circle rad. same perim.)

(17) h-f rule correction magnitude  = 0.0000002159170900 (mult. const = 1)

(18) MVR proton field correction  = 0.0000002159927198

(19) expt correction magnitude    = 0.0000002159927198

(20)   MVR high = 3288087212.2131 (MVR f-s without correction)

(21)   MVR high = 3288087922.1673 (MVR with rule)

(22) -> MVR high = 3288087922.4160 (MVR spin-2 proton field) <-

(23) -> expthigh = 3288087922.4160 (Horbatsch and Hessels, 2016) <-


PARAMETERS SEARCHED FOR ALIGNMENT

(24) -> Kalpha f-s multiplier =   1.0000000356165786 <-

(25) -> spin-2 eccentricity   =   0.9116435022877840 <-


## D.3.3  *Example 3: $2P_{1/2}$ high hyperfine, blue shift*

Level: 2P1/2 Sense: - Rotation: 135 Doppler shift: blue

Quantum numbers -> QM: n = 2  L = 1  Spin = -0.5; MVR: na = 1  nr = 1


FINE STRUCTURE ENERGIES (MHz, Kalpha = 1)

(1) Sommerfeld f-s =  822026487.4329

(2) MVR f-s       =  822026485.9391


COORDINATES (multiplied by ground state radius)

Perigee (Kalpha = 1)

(3) X =    7.9821706569 (p-e ring separation)

(4) Y =    0.5358972812 (ring radius)

(5) X*X + Y*Y =   64.0016508702 (radius^2 from proton)

Apogee (Kalpha = 1)

(3) X =    2.8799014074 (p-e ring separation)

(4) Y =    7.4643157342 (ring radius)

(5) X*X + Y*Y =   64.0017114056 (radius^2 from proton)

Collapse to 1S1/2 radius (adjusted Kalpha)

(6) X =    1.7324126792 (p-e ring separation)

(7) Y =    1.0000000000 (ring radius)

(8) X*X + Y*Y =    4.0001647538 (radius^2 from proton)


DOPPLER VALUES (at 1S1/2 radius)

(9)  cosine = 0.8661885013274804 (cosine of emission angle)

(10) Dopp factor = 1.0000012171057252 (blue shifted emission)


PRE-EMISSION (before Doppler effect)

(11) low h-f/Dopp   = 822025486.4745  high h-f/Dopp  = 822025545.6439

(12) MVR h-f mid-pt =  822025516.0592 (Kalpha: alignment h-f energies mid-pt)

(13) expt h-f mid-pt =  822025516.0592 (pre-Doppler h-f energies mid-point)

POST-EMISSION (after Doppler effect)
(14) MVRDopp     =  822026516.5512
(15) expt h-f mid-pt =  822026516.5512

HYPERFINE ENERGIES
(16) spin-2 major axis =  1.5107411982684239 (x spin-2 circle rad. same perim.)
(17) h-f rule correction magnitude  =  0.0000000359857026 (mult. const = 1/3)
(18) MVR proton field correction   =  0.0000000359900191
(19) expt correction magnitude     =  0.0000000359900191
(20)   MVR high =  822026516.5512 (MVR f-s without correction)
(21)   MVR high =  822026546.1324 (MVR with rule)
(22) -> MVR high =  822026546.1359 (MVR spin-2 proton field) <-
(23) -> expt high =  822026546.1359 (Horbatsch and Hessels, 2016) <-

PARAMETERS SEARCHED FOR ALIGNMENT
(24) -> Kalpha f-s multiplier =  0.9999994100773449 <-
(25) -> spin-2 eccentricity  =  0.9849763740076631 <-

### D.3.4   *Example 4: $3P_{3/2}$ high hyperfine, blue shift*

Level: 3P3/2 Sense: + Rotation: 135 Doppler shift: blue
Quantum numbers -> QM: n = 3  L = 1  Spin = 0.5; MVR: na = 2  nr = 1

FINE STRUCTURE ENERGIES (MHz, Kalpha = 1)
(1) Sommerfeld f-s =  365340646.9762
(2) MVR f-s     =  365340646.6816

COORDINATES (multiplied by ground state radius)
Perigee (Kalpha = 1)
(3) X =   17.8547170946 (p-e ring separation)
(4) Y =    2.2918858660 (ring radius)
(5) X*X + Y*Y =  324.0411672487 (radius^2 from proton)
Apogee (Kalpha = 1)
(3) X =    8.7913808911 (p-e ring separation)
(4) Y =   15.7089129467 (ring radius)
(5) X*X + Y*Y =  324.0412135541 (radius^2 from proton)
Collapse to 1S1/2 radius (adjusted Kalpha)
(6) X =    1.7324126792 (p-e ring separation)
(7) Y =    1.0000000000 (ring radius)
(8) X*X + Y*Y =    4.0001647538 (radius^2 from proton)

DOPPLER VALUES (at 1S1/2 radius)
(9) cosine = 0.8661885013274804 (cosine of emission angle)
(10) Dopp factor = 1.0000008113980976 (blue shifted emission)

PRE-EMISSION (before Doppler effect)
(11) low h-f/Dopp = 365340344.1675 high h-f/Dopp = 365340351.1754
(12) MVR h-f mid-pt = 365340347.6715 (Kalpha: alignment h-f energies mid-pt)
(13) expt h-f mid-pt = 365340347.6715 (pre-Doppler h-f energies mid-point)

POST-EMISSION (after Doppler effect)
(14) MVR Dopp = 365340644.1080
(15) expt h-f mid-pt = 365340644.1080

HYPERFINE ENERGIES
(16) spin-2 major axis = 1.5414034604555343 (x spin-2 circle rad. same perim.)
(17) h-f rule correction magnitude = 0.0000000095961022 (mult. const = 2/15)
(18) MVR proton field correction = -0.0000000095909121
(19) expt correction magnitude = 0.0000000095909121
(20) MVR low = 365340644.1080 (MVR f-s without correction)
(21) MVR low = 365340640.6021 (MVR with rule)
(22) -> MVR low = 365340640.6040 (MVR spin-2 proton field) <-
(23) -> expt low = 365340640.6040 (Horbatsch and Hessels, 2016) <-

PARAMETERS SEARCHED FOR ALIGNMENT
(24) -> Kalpha f-s multiplier = 0.9999995907807657 <-
(25) -> spin-2 eccentricity = 0.9938223386951067 <-

# Reference

Mohr, P. J., D. B. Newell, B. N. Taylor, and E. Tiesinga. 'Data and analysis for the CODATA 2017 special fundamental constants adjustment'. *Metrologia*, 55 (2018): 125–146.

# Index

# About the Author

 **Dr. Barry R. Clarke** has a variety of interests. He has academic publications in quantum mechanics and Shakespeare studies, and has published books of original mathematics and logic puzzles for Cambridge University Press and Dover Publications. The present work is a sequel to *The Quantum Puzzle: Critique of Quantum theory and Electrodynamics*. His book *Challenging Logic Puzzles Mensa* has sold over 100,000 copies, his academic text *Francis Bacon's Contribution to Shakespeare: A New Attribution Method* has attracted national media attention, and he is presently puzzle compiler for *The Daily Telegraph* and *Prospect* magazine (UK). His puzzle work and his comedy sketches have been broadcast in the UK on both BBC and ITV television.

Printed in the United States
by Baker & Taylor Publisher Services